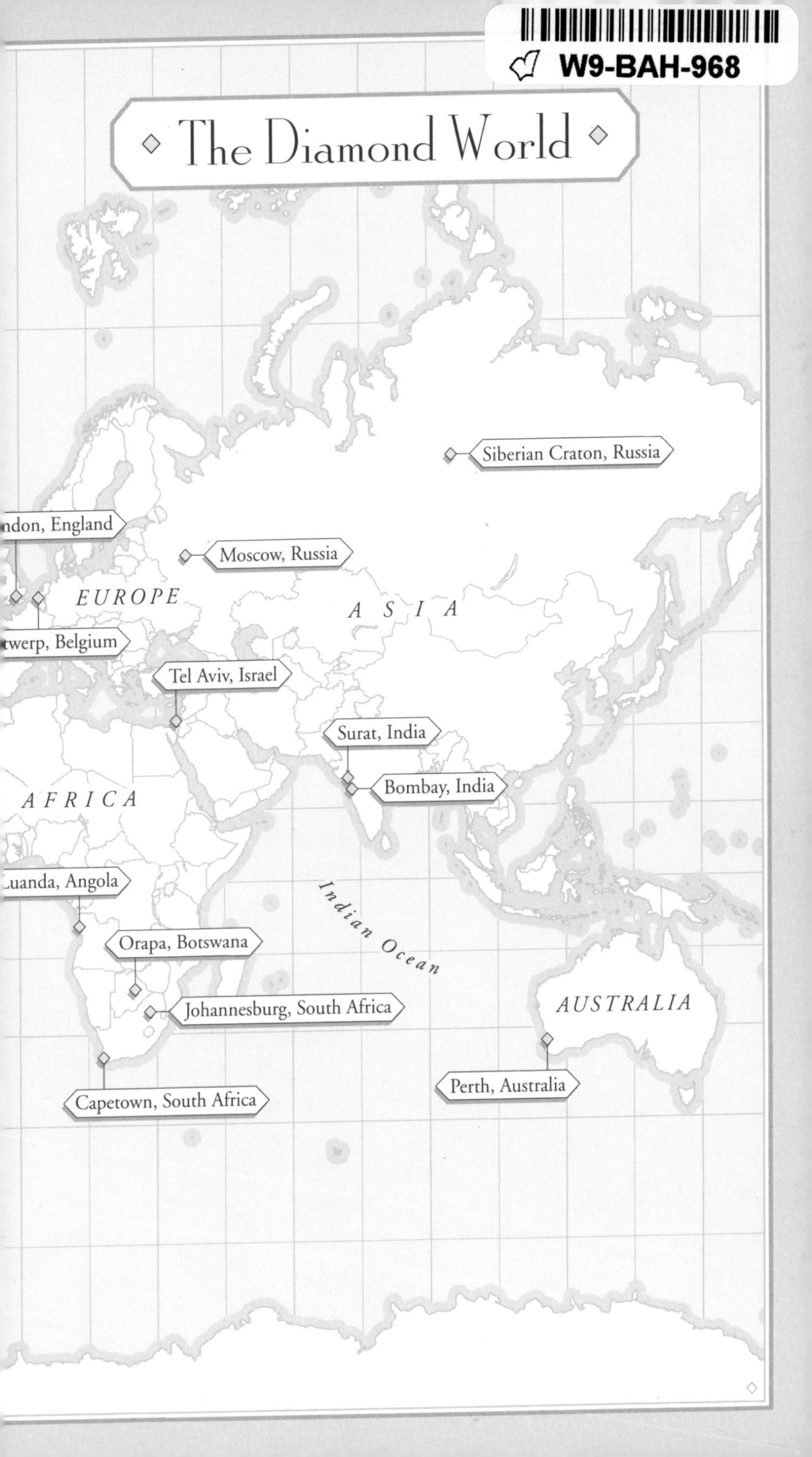

The Diamond World
Siberian Craton, Russia
ndon, England
Moscow, Russia
EUROPE
A S I A
twerp, Belgium
Tel Aviv, Israel
Surat, India
Bombay, India
AFRICA
Luanda, Angola
Indian Ocean
Orapa, Botswana
Johannesburg, South Africa
AUSTRALIA
Capetown, South Africa
Perth, Australia

Diamond

A Journey to the Heart of an Obsession

WALKER & COMPANY
NEW YORK

Diamond

◆

Matthew Hart

First published in the United States of America in 2001 by
Walker Publishing Company, Inc.

Library of Congress Cataloging-in-Publication Data

Hart, Matthew, 1945–
Diamond: a journey to the heart of an obsession / Matthew Hart.
p. cm.
Includes bibliographical references.
ISBN 0-8027-1368-8 (alk. paper)
1. Diamonds. I. Title.

TS753.H37 2001
553.8'2—dc21 2001026348

Book design by Ellen Cipriano

Printed in the United States of America

2 4 6 8 10 9 7 5 3 1

To Heather

CONTENTS

1. A Large Pink 1

2. The Diamond Seas 22

3. The Middle Empire 32

4. The Long Hunt 60

5. Rush in the Barrens 84

6. The End of the Old Cartel 113

7. The Manufacture of Desire 138

8. Bad Goods 159

9. The Diamond Wars 182

10. Diamond Cutter 201

11. Rosy Blue 223

12. The Dogrib Country 241

Appendix 253

Selected Bibliography 255

Acknowledgments 259

Index 263

Diamond

1 A Large Pink

On the morning of May 9, 1999, on the upper reaches of the Rio Abaete in the state of Minas Gerais, Brazil, three garimpeiros anchored their barge in the muddy current and began to suction gravel from the river. A *garimpeiro* is a small-scale miner; the area he works is called a *garimpo.* It was a Sunday, usually a day of rest. But the pickings had been poor, and the men needed money, and so they had started their outboard engine and chugged upriver from their camp and put their lines ashore to hold the barge—a platform on pontoons—against the current. An awning sheltered them from the hot sun.

The miners sucked up a cubic yard of material from the riverbed, washed away the muck, and began to sieve the remaining gravel. They worked steadily for two hours without luck, until a thin stone caught in the screen. It was roughly triangular in shape and measured about an inch and a half on the longest side. The garimpeiro working the screen shouted, and waved at his companions to shut off the pump. He washed the stone and rubbed it with his fingers and took it in a cloth and wiped it clean. With few words they passed the stone around. Each man held it to the sunlight, squinted, and agreed that, yes, the stone was definitely pink. They hurriedly cast off their lines and headed the barge to shore. It took them an hour to get to a phone and place a call to Gilmar Campos, who owned the barge they worked on and was therefore a majority owner of the stone.

Because it was Sunday, the Campos brothers were not in their office in Patos de Minas, eighty miles west of the garimpo on the Abaete. Patos de Minas is a city of 150,000 in western Minas Gerais, an area of historic diamond interest. The Campos brothers operate a truck-parts business in a building where they also trade diamonds. The building faces a man-made lagoon, and on a Sunday the citizens of Patos stroll about with their families on paths that border the sluggish water. On an island in the middle of the pond, cormorants hunch like dispirited monks. In this uneventful Brazilian backwater the truck-parts business fits in. Large, painted letters on the side of the building describe the automotive enterprise. No sign proclaims that the Campos brothers buy and sell rough diamonds worth millions of dollars a year, and that at any time they may have a small pink, or a fabulous green, or a good, clear, 50-carat white sitting up there on the second floor, above the brake pads and the carburetors.

Gilmar Campos, the eldest brother and head of the family, is a hard man in his forties, with an olive complexion and protuberant eyes and thick, black hair. The youngest brother, Geraldo, athletic-looking and restless, is the head of the truck-parts business. Between the other two in age and station is Gisnei, a former civil servant with an easygoing manner, the reputed peacemaker of the family, the atomic bond that allows the strong personalities of Gilmar and Geraldo to combine.

"Put down that Gilmar saw it first," Gisnei said of that Sunday in May when the garimpeiros called. "Put down that Gilmar got to the Abaete first, and was the first to see the stone." In fact Geraldo got there first, since he was the one who had his cell phone with him when the garimpeiros phoned. But the order of arrival was a matter of pride, given subsequent events, and Gis-

Garimpeiros' barges on the Abaete. (Matthew Hart)

nei, soother of egos, wanted to make sure that Gilmar got the credit.

As they sped east out of Patos in their cars, each of the brothers was thinking the same thing: Was the stone a real pink? The garimpeiros had claimed it was. Most likely, as the brothers knew, it would turn out not to be. A true pink diamond is very rare, and although Brazil is a good source of such stones, the odds were that this one would be a more common brown color, with merely a pinkish tint. Such false pinks will not survive the process of polishing, and will fade into brown. Even so, the stone had another attribute that by itself would have sent the Campos brothers racing out onto the highway and tearing eastward through the countryside—size.

The garimpeiros had reported the stone at 81 carats. That weight would be accurate, since the garimpeiros had a diamond scale. If the stone were a true pink as well, it would be worth millions. The brothers gunned their cars recklessly as they turned off the paved highway and onto the red dirt road that led to the river. Gisnei recalled that he was shaking so much he could barely drive. Geraldo arrived first. He got out of his car and the garimpeiros came to meet him. They handed him the stone. He fished out his loupe and studied it, then looked away to clear his head and took a deep breath and looked again. It was a strong pink; he had never seen its like. "I felt great emotion," he would later say. "My feelings were very great." When Gilmar got there, Geraldo handed him the stone. Gilmar, the hard man, took one look and began to cry.

When the emotions of the moment passed, the brothers thought about the implications of the stone. Gilmar knew he needed help. For one thing, the garimpeiros were insisting the stone was worth millions. They were adamant about it. Since Gilmar owned the barge and the pump, he would have the lion's share of the pink. But the garimpeiros' cut would be one fifth, and they wanted their money quickly. Gilmar decided to bring in a partner, not only to buy out the garimpeiros' interest but to help establish the value of the stone and to sell it.

Large, colored diamonds are difficult to value. Each is unique. They do not conform to standard pricing conventions of the diamond trade. They are worth what the seller can get for them. The number of potential purchasers is small. The introduction of a newly discovered diamond into this rarefied world is a matter of extraordinary delicacy and bluff. Gilmar Campos did not know where to begin, but he knew someone who did.

He called Luigi Giglio, a diamond dealer and miner based in Coromandel, eighty miles west of Patos de Minas. The call to Giglio also brought in Stephen Fabian, Giglio's associate. Fabian, an Australian, was a mining engineer and mining-stock analyst who had masterminded a deal that rolled Giglio's Brazilian mining prospects into a company called Black Swan Resources. Fabian had remained in Brazil to run Black Swan.

Giglio was a powerful figure in the world of Brazilian diamonds. Of medium height, with a characteristically stubbled face, the chain-smoking Giglio could easily have been mistaken

Gilmar Campos with the pink. (Matthew Hart)

for a small player, hunched over a beer late at night in a backstreet café in Coromandel, his arms smudged with the dirt of some diamond river. He began trading rough stones while still in his teens. At the age of twenty-three he settled in Coromandel, where he lived at the time of these events, occupying alone the large, walled, ocher villa that was the largest house in town. He ran a thriving diamond-trading business, and a polishing factory, and kept a team of miners busy in the shoreside gravels of a nearby river. Every few months he flew out to Rio de Janeiro to visit his wife and daughters, stayed for a day or two, and then hurried back to Coromandel to continue his restless search for diamonds.

Brazil was once the world's leading diamond producer. Although its glory days are past, the country retains a warm place in the hearts of diamond traders for the quality of its gems. Brazil's diamonds come from rivers. Such diamonds, called "alluvials," are on average better than stones mined from open pits. The Abaete once yielded an 827-carat stone. Pinks weighing 275 carats and 120 carats, respectively, have come out of the muddy stream. In 1938, garimpeiros on the Rio Santo Antonio do Brito found the 727-carat Presidente Varga, the most famous diamond in Brazil. The same river has produced rough diamonds weighing 602 carats, 460 carats, 400 carats, and 375 carats. An amazing parade of gems has marched out of the diamond valleys in the last 250 years, and some of the more recent have arrived at Giglio's door in the middle of the night, wrapped in rags by excited garimpeiros.

Garimpeiros occupy a romantic place in the public imagination of Brazil, a country with many poor. The miners' rights derive from centuries of practice, rather than from law, and a fine air of sanctioned banditry burnishes their popular image. A garimpeiro may live in a riverside hovel roofed with plastic, or own a house in town. He may drive a car or go on foot. But regardless of his fortune, one rule never varies: when he finds a diamond, someone pays him for it. In the case of the pink, it did not matter that the barge belonged to Gilmar Campos. The garimpeiros had found the stone; they were entitled to their cut; they must be promptly paid. What is more, they had hit on a price—$2 million—and would not be budged.

The Campos brothers showed the pink to Giglio. Certainly he was captivated by it—the color was a good, strong pink. The discussions went this way and that. Giglio and the brothers considered an unsightly protuberance at one corner, and agreed it should be cleaved away to improve the appearance. This reduced the weight by two carats. Giglio thought the stone looked very good, and suggested to Fabian that Black Swan make the required investment.

Fabian did not immediately see the merit of this proposal. Although Giglio had a large block of shares, Black Swan was a public company, listed in Toronto. It had been formed to search for diamond deposits, not to speculate in rare diamonds. Still, Fabian saw that the story of a large and valuable diamond would attract publicity, raising the company's profile with investors. He agreed that Black Swan would buy a share of the pink, with two provisos: Giglio must guarantee a minimum return to the company, and Black Swan must secure an outside valuation. This second stipulation set in motion events that would move the news of the discovery from the small group that held it into the wider diamond world.

A garimpeiro. (Matthew Hart)

After a series of phone calls and referrals, Fabian found his way to Richard Wake-Walker, a London diamond consultant. Wake-Walker understood the challenge right away: any assessment would hang on color. Was the diamond truly pink? Would the color survive the cutting and polishing of the stone, or would it fade, or perhaps improve? The answers to these questions would be guesses, since the final color of a diamond can never be known until the stone is finished. But there are people competent to make such guesses, and the man Wake-Walker wanted was about to leave London to return to South Africa. Wake-Walker placed a rapid call and managed to intercept Mervin Lifshitz, a Johannesburg diamantaire.

"Diamantaire," the diamond trade's term for dealer, implies more than commercial acumen. It describes expertness, sagacity, and, at its best, a sympathetic aptitude for diamonds. When Wake-Walker reached him and told him about the pink, Lifshitz was interested at once. He agreed to go to Brazil and take a look at the diamond. But he was pressed for time: business demanded his quick return to South Africa. Moreover, he happened to be without the equipment he would need for such an important examination. So while Wake-Walker's travel agent got onto the computer to work out flights, Lifshitz rushed in a cab to London's diamond quarter, in Hatton Garden, where he bought a loupe, a portable lamp, and a measuring gauge. Then he headed to the airport.

Lifshitz flew to Zurich and caught the overnight flight to São Paulo. From São Paulo he flew to Belo Horizonte, where Fabian waited with a chartered turboprop to take them to Monte Carmelo. Giglio met them there, shook hands with Lifshitz, and the three drove straight to Giglio's polishing factory in Coromandel, where the stone was locked in a safe. Lifshitz waited in an office while Giglio got the diamond and brought it to the room. Twenty hours after leaving London, Lifshitz unfolded the paper packet.

"My first reaction was—it took my breath away," said Lifshitz. "I opened the paper and it freaked me out. I closed it and put it back on a table and walked around the office and then came back. The first impression is that what you are seeing and your mind are not in sync. It's gorgeous. The stone is just gorgeous. Then the second time [he opened the packet] is when I sat down and made an analysis. I spent about three hours with the stone."

When Lifshitz completed his examination, the three men drove to Giglio's house. Over lunch, Lifshitz told them that the pink was worth a lot, but he didn't want to put a price on it right away. He would review his notes later and send his report to Wake-Walker in London. After a hurried lunch they drove Lifshitz back to the airstrip and the plane took him directly to São Paulo, just in time for the South African Airways weekly flight to Johannesburg.

By the time he arrived in Johannesburg at eight o'clock the next morning, Lifshitz had been on a plane for the best part of two days, but he went straight to his office, bolted down some coffee, and set to work. In his three hours with the stone he had made extensive notes. He spread them out on his desk, set them in order, and marshaled his thoughts. The report that finally emerged contained not only Lifshitz's assessment but also Wake-Walker's thoughts on the state of the market, and in particular the market for the pink. The report on the stone ran to eleven pages. Nestled within it were six key sentences that justified Lifshitz's trip to Brazil, Giglio's keen interest, and the Campos brothers' hopes:

> At first sight the diamond strikes the examiner as being undeniably pink in color. It is my opinion that there is an absence of brown in the stone. Under normal sunlight the rough diamond shows up as being very strong pink with a very even color distribution throughout. It has a very distinct deep pink color in sunlight.
>
> The rough diamond has a cleaved surface on one face and a rough surface on the opposite. Taking this into account, the pink color is equal in intensity from both sides.

The strength of the finished color would set the price per carat of the cut stone. Hundreds of thousands of dollars per carat, therefore, rode on the answer to the question: How pink would the pink be? Lifshitz addressed this next. The worst case was that the diamond would polish to a brownish pink, and Lifshitz rated that chance at 20 percent. The best scenario was that the stone would yield a "vivid pink," and this was given a 30 percent chance. The most probable outcome, Lifshitz thought, was that the rough would produce polished jewels of a "deep pink" (lighter than "vivid," but very good), and he rated that chance at 50 percent.

Based on the color outcomes assigned by Lifshitz, Wake-Walker and his team put the pink's value at between $6 million and $20 million. They thought the best way to sell it would be at private tender, and recommended a reserve price, below which the owners would not sell the diamond. They put the reserve at $130,000 a carat, or just over $10 million. Obviously the price was conjectural. It would depend upon who wanted the stone, and how badly they wanted it. The report from Wake-Walker's London group said as much, adding that weak oil prices and the then-recent collapse of Asian stock markets had sidelined a number of important buyers, including, presumably, the oil-rich sultan of Brunei. In spite of these uncertainties, Black Swan paid $2 million to the Campos brothers for a one-sixth share of the stone.

Now Giglio and the brothers determined on a bold move. Instead of inviting buyers to come to Brazil to see the stone (a common practice with extremely rare goods), they would take the gem straight to New York, the capital city of large diamonds. They wrapped the pink in a cloth and put it in a leather pouch. Gilmar Campos strapped it under his shirt. They flew to Miami, then on to New York. They took the stone to Harry Winston on Fifth Avenue, where they were shown upstairs to the private office of Ronald Winston, son and business successor of the legendary Harry. Winston sent for his chief polisher and one of his veteran salesmen. Gilmar then unwrapped the stone and tossed it onto Winston's blotter. There was a little hush while Winston and his people stared at it. A 79-carat rough pink diamond is worth a drawn breath anywhere, even at Winston's.

There were guarded murmurs of admiration. The brothers said, yes, it was a fantastic stone, and the Americans could have it for $20 million. Winston said he'd like to think about it. No problem, said Giglio; they took their stone and left, and went out that night and tore up the town, partying into the morning, the big pink diamond nestled in its pouch as they wandered from bar to bar.

Next day they went back to the Fifth Avenue store. Winston's cutter said he'd like to polish a window onto the stone to take a look inside. A window is a facet polished onto a piece of rough, as uncut diamonds are called, to enable a tactician to assess the interior of a stone before cutting it. In this case, what they wanted to check was the color. Giglio and the brothers refused. Putting in a window, as one of them would later say, would have been like having sex before the Americans had even asked them out on a date. They decided not to stay in New York or show the stone to other dealers, because it might be interpreted as anxiety to sell. They returned to Brazil.

News of the pink spread quickly, as they had known it would, and soon William Goldberg paid them a visit. Goldberg, a New York diamantaire, is an extravagant character with long white hair and a husky voice, like Marlon Brando's in *The Godfather.* Goldberg had led the syndicate that bought and cut the Premier Rose, a top-color white diamond that weighed 353.9 carats in the rough. It was Goldberg too who had polished the famous Pumpkin diamond, a vivid orange with a finished weight of 5.54 carats, bought by Harry Winston for $1.3 million. Large stones are notoriously treacherous, and few diamond manufacturers have the resources or the nerve to attempt to polish them.

The brothers met Goldberg at the airport and drove him to the truck-parts shop. The old dealer mounted the stairs to the second floor, wheezing with the effort, and sat himself at Gilmar Campos's desk. They brought out the pink and Goldberg examined it. He offered them $3 million. The brothers said no. Goldberg thought the diamond was too risky for a higher offer, and returned to New York.

Stones like the pink excite the romantic instincts of the diamond trade, and its gambler's heart, and reveal the sense of breathlessness and quest that attends the diamond game. I heard about the pink myself in August of that year, when a trader in polished goods phoned me from New York.

"Did you hear about that big red diamond that somebody's got?" he asked.

"Red?" I replied.

"Yes, red. Very very pink. Red."

"A blood red?"

"It's very very red. Some guy from, like, Ohio, brought it into town. He inherited it. He had no idea what it was, and took it around to see what it was worth. He showed it around the street. Some counter guy looked at it and freaked."

That the Campos brothers' pink had shot into this stratosphere of rumor was not surprising. Great diamonds gather such misinformation all the time, and it does not dull their glitter. A little harmless balderdash is part of the proper business of the trade.

The seller of a rough diamond wants to leave plenty of room for speculation. As Wake-Walker had said, "Sometimes the greater the discretion that such a stone can be handled with adds to its value. It is not necessarily helpful for the vendor of the polished if the whole world knows how much the rough cost." Giglio and the brothers followed this advice. Details about the pink were hoarded frugally. The diamond street supplied from its own imagination whatever it could not learn. Rumor grew on rumor in a steady and inevitable seduction. I became snared in this myself, and determined to track down the diamond, and by the time I had located Fabian and obtained a copy of Wake-Walker's report, infatuation had me in her arms. I had to see the pink before it was cut.

◆

A rough diamond is pure potential and pure risk. No one can know for certain what will happen when the stone goes onto the polishing wheel or when the sawyer puts it to the saw. Diamonds contain flaws that evade detection, even by technical means. This unknowableness accounts for the passion diamantaires have for rough. The most hard-bitten fanciers of rough, those who are hooked by the sheer hazard of an uncut stone, will tell anyone who stops within earshot that a polished diamond is just ruined rough.

The shape of a rough diamond is an important factor in deciding the shape of the polished stone. Mervin Lifshitz believed that a pear would be the "natural" shape for the pink. If they polished the pink into a pear, the yield of the rough would be 50 percent. In other words, the polished diamond would be half the weight of the rough, or almost 40 carats. A polished yield of 50 percent is an excellent return from rough. In the case of the pink, however, as Lifshitz knew, a large pear would be possible only if the rough had no major flaw. In fact it did.

An assortment of rough diamonds at the Diamond Trading Company, London. Each pile is valued at $500,000, demonstrating the wide range in per-carat prices between large and small stones. (De Beers)

The diamantaire had detected a gletz, or tiny crack, that ran through half the length of the rough. The gletz put the whole stone at risk. If a cutter attempted to grind away part of the gem, the stone could shatter into fragments. The diamond would have to be studied over many days before it was cut. The cutter would make a number of plastic models and experiment on them. Lifshitz outlined in detail four polishing options of his own.

In one option the cutter would begin by cleaving the stone in half along the gletz. Of the two resulting pieces, one would be slightly larger. The larger piece would polish into two separate stones—a 13-carat pear and a 2-carat round shape. The smaller piece of rough would have two small gletzes remaining in it, but Lifshitz did not think they would cause the polisher much difficulty. This smaller stone could polish into a 15-carat heart shape, "although it might be a bit fattish." Overall, this first option would deliver polished diamonds weighing some 30 carats, for a yield on the rough of almost 40 percent.

Each three polishing option had ramifications crucial to calculating what the diamond was worth. It would not simply be a matter of adding up how many carats each option would deliver, and picking the one with the biggest total yield. Large polished diamonds are worth more *per carat* than small ones. A 2-carat stone of a given quality, for example, sells for more than two 1-carat stones of the same quality. Such considerations would be central to anyone thinking of buying the pink in the rough.

A bewildering range of prices complicated the picture. A colored diamond, or "fancy," is classified according to fine gradations of color, and Lifshitz listed a range of prices fetched by pinks. A 7-carat fancy light pink had gone for $113,000 a carat, while a 3-carat fancy intense purplish pink had earned $260,000 a carat. These were midrange prices. Other pinks had gone for as little as $16,000 a carat and as much as $730,000 a carat. The stakes were staggering, which is one of the reasons I wanted so much to see the pink as rough.

I badgered Stephen Fabian every week, and finally he agreed to ask the Campos brothers to show me the stone. They promised nothing, but did not say no, and on January 16, 2000, I flew from Toronto to São Paulo, then on to Belo Horizonte, where I stayed the night. Fabian picked me up in the morning and we drove out of the city, heading for Patos de Minas. The sun shone brightly down and the landscape gleamed from the previous night's rain. Brazil's highways follow the highest ground to avoid flooding in the valleys during the torrential summer rains. From this lofty elevation, mile after mile of vistas stretched away. We stopped at a highway restaurant for coffee and *pão de queijo*, hot rolls baked with cheese. Then we climbed back into the truck and took the road north to the diamond country.

Diamond traders in Minas Gerais live dangerous lives. Dealers compete ferociously for stones, and *pistoleros* sometimes help to tilt the commerce one way or another. About ten thousand garimpeiros work the region, and a steady production of valuable goods comes out of the diamond rivers. The business is a cash business. Players like Giglio have ready access to large sums, a fact widely known. Giglio's diamond inventory might at any time run into the tens of millions of dollars. He has reputedly dodged several murderous assaults.

Cheats too populate the local diamond trade; every kind of fake shows up. Not long after the garimpeiros found the big pink, a trader appeared at Giglio's house late at night to offer him a pink weighing 30 carats. Giglio did not even get out his loupe. "Either this stone is stolen," he said, "and you can put it back up your ass where it came from, or it's a piece of quartz with some nail polish, and you should take it to Tel Aviv and ask for a hundred and fifty thousand dollars." Of course, sometimes even colossal stones are real. In 1997 a garimpeiro brought Giglio a 350-carat white diamond from the Paranaiba River. A rival diamond buyer had apparently offered $8 million for the stone. Giglio examined it and put together a syndicate that paid $12 million, or so the story goes.

Four hours after we had left Belo Horizonte, Fabian used his cell phone to call Giglio and ask if he had been able to reach the brothers in the last few hours, and whether they were expecting us and would show me the pink. As far as Giglio knew, they would, but it was up to them. "It will be Gilmar's call," Fabian said when he clicked off the phone. "It's just going to depend on how he feels about it right now."

We drove into Patos de Minas, through town, and arrived at the truck-parts shop beside the stagnant lake. A gate topped with barbed wire blocked the lane. Fabian drove back to a disused lot, parked, and made another call. Five minutes later Gisnei Campos appeared, strolling up the sidewalk with his arms hanging loosely at his sides. He greeted Fabian warmly, shook hands with me, and chatted in Portuguese with Fabian. All the while he shot sidelong glances at me, a bit shyly. Fabian opened the back of his truck and tugged out a box of black polo shirts with the Black Swan logo stitched in white. He added a few Black Swan baseball caps, and Gisnei grinned widely and made a joke and walked away.

"What's happening?" I asked.

"I don't know," Fabian replied, "he said to stay here. I think he was checking you out."

Ten minutes later Gisnei came back and we all walked around a corner to a three-story residential building, like a small apartment block, where we met Geraldo. He wore a white tank top, track pants, and running shoes. He and Fabian slapped each other on the back and he looked me over candidly. Gisnei opened a locked steel gate, we went through, and he locked the gate behind us. Then we climbed a marble staircase to the second floor and entered a sparsely furnished room. Gisnei and Geraldo held a whispered conference in the kitchen, and Geraldo left. Gisnei sat down on the sofa and drank a can of Coke and said nothing. Fabian and I went onto the balcony and looked at a large building going up next door, a new house for Gilmar.

"Where did Geraldo go?" I said.

"He's getting the stone. They move it around quite a bit."

Five minutes later the steel gate clanged below, followed by the sound of shoes squeaking as someone came fast up the marble steps. Geraldo came swiftly into the room, carrying a suede pouch. From the pouch he removed a folded paper, and put it on the table and opened it. There lay the pink. Fabian and I moved to the table and stared at the diamond. Geraldo placed a loupe beside the stone and stepped back. For a diamond that had excited so much speculation it seemed amazingly delicate and frail—a thin triangle of rough not two inches long, less than an ounce, frosted on one side. Gisnei watched me from the sofa. I looked at Fabian. He shrugged, so I sat down and gingerly took the pink between my fingers. I put the loupe to my eye and brought the pink up close to the lens, and took my first look at this diamond I had come so far to see.

A rough diamond viewed through a loupe presents an amazing miniature world, lucid in every particular. It is like examining a landscape made of crystal. The eye is not confounded and dazzled, as is the case with a polished jewel. The color of the diamond seemed completely tranquil. It was an unequivocal pink. In olden times, a jewel fancier describing a top-color stone would say it was of the finest "water," and the pink had that liquid feeling, shimmering and fragile, as if a dot of rosy ink had been shaken into a stream and achieved this momentary pink, which would soon be gone. It seemed that such a color could never survive a diamond wheel, but would flow away through the opening of the first facet. I understood how nerve-racking it must be to attempt to polish such a diamond, and what a risk it would be to buy it, and I felt how intently the others were watching me. So much family fortune depended upon this stone, because the Campos brothers had been minor players, and the fame of the pink was already raising them to a higher level of the trade. I put it down and rubbed my eye.

"They've named it Estrela Rosa do Milênio," Fabian said, and Gisnei and Geraldo smiled broadly. Pink Star of the Millennium. If they could get the name to stick, it might add another one or two million dollars to the price. Named stones command a premium.

Geraldo and Gisnei each took a turn with the loupe, then Geraldo wrapped the diamond quickly in the paper, put it in the leather pouch, and left to return the diamond to its hiding place. His shoes made rapid squeaks as he ran down the marble stairs, and the gate clanged shut.

◆

A great gem collects great tales, adding to its status as a jewel. One of the storied gems of Brazil, the 1,680-carat Braganza stone, came out of the same river that produced the Campos brothers' pink. The Braganza was found about 1800 by three cavaliers banished into the Brazilian interior for crimes against the state. Under the terms of their expulsion they could not live in any of the capital towns of the country, or even settle themselves in civilized society, but had to remain in the wilderness. The penalty for breaking these restrictions would be immediate imprisonment. The hard sentence drove them into the loneliest regions of the country, and they came into Minas Gerais state and crossed the Rio Plata and made their way some miles north. At length they came to the turbid stream of the Abaete, their destination. Diamonds had already been discovered in Minas Gerais, and the three had pinned their hopes on finding some. If they did find gems, they reasoned, perhaps they could win a pardon.

They searched for five years. With only rudimentary equipment, they restricted their trenching to the riverbanks, digging here and there where they thought diamonds might have caught. In the sixth year there was a drought, and the Abaete dried to a trickle. Now the riverbed lay bare; the three convicts could reach the best gravels. They constructed a dam and started washing gravel, that is when they found the stone. In the awful heat, among swarms of flies, suddenly they possessed a fist-sized diamond. But the discovery faced them with a dilemma, for they had no mining license. The stone was therefore illegal. In effect, they had stolen it from the crown of Portugal, Brazil's colonial master. Not only would they be breaking their sentence by approaching the authorities, they would be coming as thieves who had robbed the king. In the end they had no choice, and decided to put their trust in the diamond itself. The regional governor, agog at the size of the rough jewel, commuted their sentences on the spot. The diamond went to Rio de Janeiro, where the authorities put it on a frigate and sent it to Lisbon.

Braganza was the name of the Portuguese royal house, and the gem had been named for them. There has since been controversy over whether the stone was truly a diamond. If it was, it would have been the largest diamond in the world at the time, and the second largest still today, and so where is it? There is no such diamond around, or if there is, someone is hiding it. Some researchers suggest that the Braganza was not a diamond, but a white topaz, and that the royal household maintained the fiction of a diamond because they thought it sounded better. Or the Braganza might indeed have been a diamond, and been stolen in the confusion of the early nineteenth century, as Napoleon's armies struggled in Spain and Portugal with the Duke of Wellington, and the Portuguese court fled to Brazil. When the English forced the French general, the duke of Abrantes, to retreat from Portugal, the Frenchman sent a casket stuffed with forty thousand Portuguese gold coins to his wife in France. The suspicion arose that the French duke had gotten hold of the Braganza, too, and put it in with the coins.

Another scenario is that the name Braganza never did refer to the large stone found by the cavaliers, but properly belonged to a smaller diamond. There are candidates weighing 144 and 215 carats, respectively, and the rivers around Patos de Minas and Coromandel have produced an abundance of diamonds in that range. Any diamond that big would attract attention, and probably a name, before it went out into the world to wreak its havoc.

A 79-carat pink is good enough for a story too, and Estrela Rosa do Milênio did not take long to get one. It suddenly vanished from the diamond radar. Nothing could be learned about it. I made many futile calls myself. Black Swan, the only partner in the stone with public accountability (as a publicly traded company), refused to comment on the diamond's fate. In New York, where most important diamonds are cut, no one had any news of the stone. A rumor surfaced that the diamond was in Tel Aviv, where a syndicate of cutters was examining it with a view to buying a share.

The gravest risk to potential buyers of the pink lay in the chance that the color would disappear when the stone was cut. Brian Menell, a Johannesburg diamantaire and son of a distinguished South African mining family, once bought a valuable blue diamond. "It was a good strong blue," he said. "We started polishing, putting on facets. Suddenly, as the cutter added a facet, the color changed from strong blue to light blue, from $260,000 a carat to $40,000 a carat." Menell had been aiming for a 6-carat finished stone, and had therefore watched $1.3 million evaporate before his eyes. As it turned out, Menell was lucky. When his polisher put on the next facet, the color flowed back into the stone. The news of such incidents spreads quickly through the diamond world, and sows fear in the hearts of cutters who face colored diamonds.

This apprehension would explain the difficulty in selling the pink, if they were having difficulty. Or possibly they had not had difficulty, and had already sold it, and the buyer had wanted nothing said, to spare himself the embarrassment of a ruined stone, if it were ruined. If, if, if. There was nothing in the wind but conjecture. Black Swan's $2 million investment had secured a one-sixth interest in the diamond, and Giglio had guaranteed the company a 10 percent return on its money. If a one-sixth share was therefore worth $2.2 million, the price tag for the pink should be $13.2 million. But no word on this or anything else to do with the pink could be wrung from its owners. That is how matters stood for three months, until May 30, 2000, when a notice appeared on Black Swan's Web site.

> Black Swan Resources Ltd. ("Black Swan") understands that the partnership in which it invested US $2,000,000 in 1999 to acquire an interest in and market an exceptional 79-carat pink diamond, has now sold the diamond. Black Swan understands that it will receive an amount of US $2,200,000, in installments within the next two months. Black Swan is currently seeking further information and verification of facts relative to this transaction. Mr. L. Giglio, a director of Black Swan and its largest shareholder, has offered to provide security for the US $2,200,000 due to Black Swan.

I called Stephen Fabian, Black Swan's president, right away. "I don't quite get it," I said. "Did you sell it?"

"I think the brothers sold it," Fabian replied.

"Think?"

"I can't say anything about it, mate. Sorry."

"Well, where did they sell it? New York?"

"I don't really know. I can't say anything about it."

One month later a second notice appeared on the site, advising that the Campos brothers were disputing that they owed Black Swan anything at all. In August, documents filed in Toronto with the Ontario Securities Commission revealed that Black Swan was moving to seize a block of Luigi Giglio's stock, which he had pledged to cover Black Swan's $2 million investment and the promised $200,000 profit.

The pink had vanished. I could not find anyone who knew what had happened to it. Fabian finally relented and sent me an e-mail. "The stone did not go to New York," he wrote. "Try looking for a large Asian construction group based in Hong Kong, which has plenty of money in China and as a sideline deals in diamonds and other precious stones. This accounts for about three percent of their business. The stone was apparently bought for the Chairman's private collection and will never be shown publicly."

The Asian group was Chow Tai Fook, a concern with important diamond interests. I opened an exchange of e-mails with them. They asked why I wanted to know about the pink. I said I had followed the diamond, as it were, from birth, and would like to see the polished. After that I never got another word.

The three garimpeiros were very rich for a short time, and have now spent all their money.

2 The Diamond Seas

Diamonds are profoundly ancient. They have existed in the universe since before the formation of the Earth or the sun. Carbon, the stuff of diamonds, is the fourth most abundant element in the solar system and, probably, in the universe. It exists in huge reservoirs in the interiors of stars. In the violent processes of stellar evolution, this carbon suffers unimaginable pressures.

In 1987 astronomers observing a supernova (or exploding star) through spectroscopy, which analyses the light radiated by different substances, identified diamonds. The spectroscopic signature of diamond is distinct from other forms of carbon, such as graphite. These minute stellar diamonds probably formed in the fantastic pressures of the superwind thrown off by the exploding star.

Diamonds abound in the universe. If more light were present, one might see the long reaches of space glittering with jewels. As our own solar system slowly coalesced, it doubtless harvested in its swirling mass great volumes of diamond. Some of this material impregnated meteorites, and some meteorites are amazingly rich in diamond. The diamond is present in minute specks, mere millionths of a millimeter in diameter, but in staggering concentrations. A meteorite may contain diamond in concentrations as high as 1,400 parts per million—three hundred times richer in proportion than the average diamond mine on Earth.

A billion years ago a barrage of meteorites pounded the infant Earth. The bombardment lasted some 400 million years. At the time, Earth's atmosphere was thin, too thin to cause the friction that today incinerates most of the objects falling through it. So some of the infinitesimal diamonds that smacked into Earth aboard the meteorites could have survived. Given the way crystals grow by adding successive layers, it may be that some of that rain of ancient diamonds falling to Earth seeded diamonds we mine today. In this model, the diamond on someone's finger might contain at its center a dot of a jewel whose antiquity goes back 10 billion years.

◆

Although the universe is abundant in diamonds, and the Earth itself encloses great seas of diamond-bearing rock, the stones are hard to find in the quantities necessary to support the development of a mine. The plucking of diamonds out of rivers and associated riverside gravels by garimpeiros and other miners supplies only a small percentage of the world's gem diamonds. Diamond rivers are merely secondary sources of the gems. The diamonds found in them have not originated in the rivers, but in primary sources deep within the Earth. They were transported to the surface by volcanic eruptions.

The ultimate diamond source is a class of extinct volcano called a pipe, stuffed with a frequently soft and crumbly gray-green rock called kimberlite, named after Kimberley, South Africa, where it was first identified. The largest known kimberlite pipe is 361 acres on the surface. Most pipes are much smaller, and some rich diamond pipes have surface expressions of only a few acres. The term "pipe" describes a slender, carrot-shaped formation. The walls of a pipe slope steeply, at an angle of about 85 degrees, and the structure plunges deeply down, narrowing into slender dykes that penetrate the Earth to a depth of a hundred miles, where diamonds come from.

The Earth has a metallic core and a thin crust. Between them lies a vast, plastic zone of rock, 2,000 miles deep, called the mantle. In the upper mantle, in zones where the temperature reaches 1,000 degrees Centigrade and the pressure climbs to 50 kilobars, carbon exists as diamond. The region where exact diamond-forming conditions prevail is called the diamond stability field, and it is only there, in all that immensity, that carbon atoms are pressed together into layers, and another layer laid upon that, until a diamond is born.

Like all crystals, diamonds grow by adding layers. With diamond, each layer is a repetition of millions of atoms, which have laced themselves together. The electronic structure of a carbon atom is such that it can form a uniquely strong bond with its fellow carbon atoms. A carbon atom has six electrons but space for ten. In the diamond form, the six-electron atom shares a single electron with each of the four atoms that surround it, and so gains a full complement of ten electrons. This sharing of electrons joins the carbon atoms in a linkage called a shared-electron bond, the strongest bond known to chemistry. The structure is said to be indomitable, a quality described by the Greek word *adamas*, from which later languages derived the words *diamond* and *adamantine*.

The eruption that produces the kimberlite volcano, or pipe, originates in the depths of the upper mantle. (Typical volcanoes, such as Mount St. Helens, originate much closer to the surface, in regions where the earth's crust is relatively thin.) In the kimberlitic eruption, a stream of gaseous rock plasma drills its way

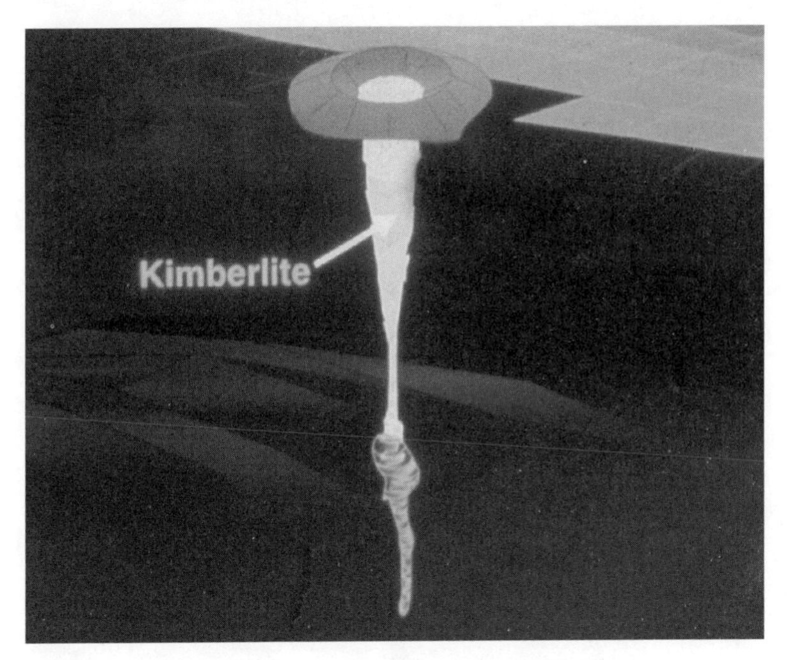

A kimberlite pipe. (BHP Diamonds)

upward, exploiting whatever weaknesses it finds in the rock above and rising at about ten miles an hour. If this rising kimberlite happens to force a passage through a zone of diamond-bearing rocks it will break some away, incorporating both rock and diamonds into the ascending kimberlite and bearing them along. A geologist would say that the kimberlite had "sampled" the diamonds. Diamonds are not products of kimberlite magma; kimberlite is merely the express elevator that takes diamonds from the mantle to the crust.

In the enormous reaches of the mantle most pipes fail to pass through diamond zones, and reach the surface barren. Even those pipes that do harvest gems from the diamond stability field may not deliver them to the surface, because diamonds in the mantle are remarkably frail. They are extremely sensitive to changes in their environment. Mantle temperature and pressure decline toward the surface, and if the rate of ascent of the kimberlite is too slow, the diamonds in it will abandon the diamond state and adopt that form of carbon that is native to such lower temperature and pressure—graphite. A rising pipe must therefore climb relatively swiftly through these diamond-hostile conditions if the gems are to survive. The hazards of the journey from mantle to crust help to account for the rarity of diamond-bearing pipes, for although the world has a known population of some six thousand pipes, only a few dozen have valuable diamond lodes.

As a kimberlite nears the surface, the pressure of the overlying rock decreases. Gases in the kimberlite expand, in the way that gases in a bottle of champagne expand when the cork is pulled. The kimberlite gushes wildly the final distance to the top, accelerating to at least a hundred miles an hour, and explodes through the surface. As it erupts, the volcano creates a swirling vortex of magma and rocks, reaming out a round shape known as a diatreme, the classic contour of a diamond pipe. The blast spews up a fountain of ejecta into the air—boulders, lava, and billions of mineral grains, including, sometimes, diamonds. If all of this just lay on the ground for anyone to find, people could go out and rummage in the bush and scrape up gemstones any time. But finding a diamond pipe is not so simple.

Considered in terms of geologic time, the surface expression of a pipe quickly disappears. The soft kimberlite breaks down, gets sluiced back into the crater, compacts, subsides, and at length is covered by the deposition of surrounding soils and glacial debris. This weathering occurs over millions of years and conceals a pipe as surely as if a hand had reached down from the sky and scumbled the earth back into place where the pipe had torn it. You could walk across the top of a kimberlite pipe and see no obvious hint of its presence. But the evidence is there, printed in the compositions of certain minerals from the diamond-bearing rocks that the kimberlite sampled. These minerals survive the weathering, and the ability to find them and recognize their significance has transformed the diamond business.

An erupting diamond pipe blows out a spray of minerals. Besides diamonds there are emerald-colored chrome diopsides and a fabulous shower of garnets. The garnets range in color from delicate pink to the deepest purple, with orange and yellow and green thrown in. These minerals and others are cousins of the diamonds, in that they are related to diamond formation, as we shall see. Because they help in the search for diamonds, such minerals are called diamond indicator minerals, or, in the shorthand of the field, indicators. The reason to look for indicators instead of for diamonds is that they are easier to find. Indicators are much more populous than diamonds. The man credited with unraveling the mysteries of the garnet, and of creating a diamond-hunting system that would threaten the supremacy of the greatest diamond miner in the world, De Beers Consolidated Mines, is a short, sunburnt geochemist from Cape Town, John Gurney.

◆

In 1970, two researchers at the geophysical laboratory of the Carnegie Institution in Washington, D.C., published a paper describing tiny bits of garnet included in diamonds. The garnets were of a type not identified before, and had a high purple color. The color came from the garnets' high chrome content. Because the garnets were included in the diamonds, the researchers suggested that these particular garnets had formed alongside the diamonds. In other words, the conditions that produced the diamonds had also produced the purple garnets. This discovery begged the question: Could a prospector find diamonds by looking for the same kind of purple garnets?

Gurney, at the time a postdoctoral fellow at the Smithsonian Institution, also in Washington, assembled samples of kimberlite from proven diamond pipes in South Africa, sorted out the garnets by color, and examined them with a microprobe, an instrument that analyses mineral chemistry. He found that some of the purple garnets had the same chemical profile as the diamond inclusions: high in chrome, low in calcium. Next he examined kimberlite from pipes known to be barren of diamonds. He found purple garnets, but not a single one that had the high-chrome, low-calcium composition.

The conclusion that diamonds could be found by locating deposits that contained these special garnets seemed inescapable. Gurney soon had a chance to test the assumption. The owners of an African diamond prospect needed an assessment of their property. Diamonds had been recovered from the prospect, and the owners were now seeking investors to help finance more work. One investor wanted reassurances about the project; and the owners asked Gurney to look at the garnets. He went through them, trying to find one that resembled the high-chrome, low-calcium garnets he believed should be present in a diamond property. He could not find a single match, and advised caution.

The owners resampled the property, and this time found no diamonds at all. A review of the exploration history revealed that the sampler on the site had always turned up diamonds in the days immediately following the visits of one of the partners. Gurney concluded that the man had been "salting" the property—seeding it with rough diamonds to distort the geological assessment, hoping to create a market value for his investment while he looked for a way to sell it. By unmasking the fraud, Gurney proved the power of the G10 garnet, as the high-chrome garnets had been classified. He had shown a clear link—no G10s, no diamonds—and saved potential investors from a multimillion-dollar trap.

Gurney published his initial research in 1973. It sounded a thunderclap throughout the mining world. Suddenly, here was a technique for finding a diamond pipe. Not only that, it was plainly posted in the public domain. Anyone could read about it. This was something new in the annals of diamond geology. The largest technical diamond establishment—that of De Beers—was famously secretive. No one knew a diamond like De Beers did, and they meant to keep it that way. The conditions were perfect for a clash between the huge miner and the independent-minded Gurney.

In the wake of Gurney's publishing bombshell, De Beers showed keen interest in his work. The company agreed to support the thesis research of one of Gurney's graduate students. The aim of the thesis was to establish the occurrence of indicators in certain diamond-bearing kimberlites in South Africa. When the thesis was published, in 1974, De Beers asked Gurney for a confidentiality agreement, citing as its reason that some of the minerals the student had examined were from De Beers mines. Gurney refused. He insisted that the knowledge did not belong to De Beers. "I told them, 'I found this out myself.'"

By insisting on autonomy, Gurney leveled an implicit threat at one of the most powerful conglomerates in the world. De Beers and its sister corporation—Anglo American Corporation of South Africa, a gold-mining powerhouse—accounted for about half South Africa's economy, measured as a share of the Johannesburg Stock Exchange's total listed capital. The Oppenheimers, who controlled De Beers and Anglo American, were the country's richest family. South Africa's banks and insurance companies, the backbone of its financial life, were enmeshed in the fortunes of the Oppenheimers' gold-and-diamond empire.

The diamond sector of the business, worth billions of dollars a year, maintained profitability by operating a cartel. It kept a tight control of supply, which gave it mastery of the price. If the price showed signs of weakening, the cartel had only to cut back the flow of diamonds to the market for the price to recover. The power of the cartel lay in managing supply. Supply was its strength, but its weakness too. If a source of goods were to appear outside control of the cartel, the diamond price—an artificial one—would quickly suffer from the competition. Gurney's G10 garnet could unlock the door to new supplies.

Gurney's passion, a scientific one, took its breath from another diamond emotion, greed. A $50-billion-a-year market exists for diamond jewelry, which supports a $6-billion-a-year demand for rough. The profits from diamond mining can be very high. There was no reason that De Beers must have all these profits, and Gurney's discoveries helped put a small army of competing diamond hunters into the field against the cartel. Even a prospector, working alone, now had an arsenal of detective weapons, for the work that Gurney started led to the development of a remarkable system for finding diamond pipes.

◆

The celebrated G10, the mineral that led the way, is of a class of garnet called pyropes, a term that derives from the ancient Greek word for fiery-eyed. Pyropes are usually deep red or purple. The high-chrome, low-calcium profile of the G10 is called its harzburgitic signature, after harzburgite, a diamond-bearing mantle rock. Harzburgite is a type of peridotite, the dominant rock of the upper mantle. Within the vast sea of peridotite are pockets of another kind of rock—eclogite. Eclogite too may be diamond bearing, and is 50 percent garnet. Diamond pipes sometimes contain eclogite boulders with a diamond content as high as 10 percent by volume—100,000 times richer than the surrounding kimberlite. The garnet for eclogite is not the G10, but an orange-colored garnet chemically distinct from its pyrope cousin. Because eclogite is so rich, it is important to know whether an exploration target may contain it, and technicians scour the soil samples for a glint of orange.

Other minerals are important too. Geologists have learned what happens to certain minerals at extremes of heat and pressure. If they find these minerals in a pipe, they can use them to decide whether the pipe has sampled the diamond stability field. Chromites, for example, need high pressure to form. They tell the explorer whether the pressure was right for diamonds in that part of the upper mantle sampled by the pipe. Chrome diopsides do the same for temperature.

Another relevant fact is that kimberlite pipes often occur in clusters. Naturally, explorers want to know how many pipes may be lurking in an exploration zone. The indicator that helps in this is ilmenite, a lovely, silvery black mineral with a subtle bluish tint. Ilmenite grains manifest a wide variety of chemical composition, distinct from pipe to pipe. No two kimberlite pipes will produce chemically identical populations of ilmenites, which means that a researcher who untangles the different compositions can predict the number of pipes in a cluster.

Most diamond-bearing pipes are found in the oldest parts of the Earth's crust, where the basement rock is more than 2.5 billion years old. The picture has emerged of deep, stable keels of rock descending to the depths where diamonds form. When an erupting diamond pipe penetrates such a structure, the tranquil rock of the keel preserves the diamonds as they ride upward in the pipe. On top of these deep keels lie the thickest parts of the continental crust—slabs of old rock called cratons. The place to look for a diamond pipe is on a craton. If you live on one yourself, there is no reason why a kimberlite should not blow up in your backyard, wrecking the flowerbed and showering the house with diamonds, but it is not likely. The youngest known diamond-bearing kimberlite pipe is 47 million years old, and even in periods of intense kimberlitic volcanism, millions of years may separate eruptions.

The diamond geologist possesses this grand model of cratonic diamond emplacement. The diamond-laden magma came up through the cool rock, rushing the last short distance and exploding through the crust. It littered the ground with diamonds and, in much greater number, pyrope garnets and other diamond cousins from the upper mantle. Prospectors had known for years that these minerals were associated with diamonds, but had not known how. After Gurney they did, and by scraping up the pretty grains and analyzing them, could lucidly extrapolate the conditions from which they came. It is a wonderful plan for finding diamond mines, and it helped upset the order of the diamond world that had prevailed for more than a hundred years.

3

The Middle Empire

Between the distant past of diamonds and the uncertain present lies a great middle era, when diamonds appeared in volumes never seen before and a succession of powerful men struggled to master them. One may look back on that time as on an age of vigorous discovery and conquest. A confluence of luck, science, and the imperial ideal spilled a tumult of men onto the plain of southern Africa. They went out in the tens of thousands and made the most furious assault on the geology of a place that the world had seen. They had great visions and petty antagonisms and when they had properly settled things among themselves, and learned the true nature of their quarry, the great age of diamonds began.

Until the last quarter of the nineteenth century, no one knew about diamond pipes. In ancient times, India was the world's diamond supplier. The Indian mines were alluvial digs. That is, the diamonds had risen from the upper mantle in pipes, and over millions of years had washed into rivers, where people found them. In the middle of the eighteenth century, Brazil supplanted India as the biggest diamond producer, and the Brazilian deposits were also alluvial. The fact that prospectors today ransack the world for pipes is due not only to recent advances in prospecting science but also to the succession of events that began in Africa, and raised the empire that shaped the modern diamond world.

In 1836, a large group of Dutch farmers in the Cape of Good Hope, embittered by British rule, loaded their life's possessions into wagons and plodded northward. They left behind the vineyards and towns and the fair valleys of the cape, and went out into a harder Africa. As they trekked north, the farmers made their way onto the Kaapvaal Craton, one of the great, imperturbable islands of rock that float upon the earth's mantle, separated by wide expanses of younger, more restive rock.

The Kaapvaal Craton is more than 2.5 billion years old. It extends north into present-day Botswana, where it underlies the Kalahari Desert. Of course, the structure of the craton, buried under millennia of sand and grit, was not visible to the farmers. Thorn trees dotted the harsh landscape. At length the trekkers came to the junction of the Orange River and the Vaal River, and between the rivers they laid out their farms.

The farmers displaced the native Griqua people, and began to scratch away at the topsoil and plant their crops. They did not notice diamonds, although the stones were there, mingled with the soil like grains of sugar in an expanse of sand. In 1859 a Griqua boy found a 5-carat stone and took it to the Berlin Mission Society at Pniel, on the Vaal. The priest who saw it must have suspected what it was, because he paid the boy £5, a lot of money. This news made its way to the cape, but nothing came of it.

Eight years later, in 1867, a young farmer named Schalk van Niekerk, needing money, started buying and selling pretty stones. Most of these came from the shallow reaches of the Vaal, where children found them. A local surveyor, who had heard about the gem found by the Griqua boy, advised van Niekerk to look out for diamonds. Van Niekerk mentioned this to a farm wife, who remembered that her son had brought home a shiny stone to use in a game. Van Niekerk found the stone in the dirt and, knowing that diamonds could scratch glass, drew it firmly across a windowpane. It left a clear mark. The pane of glass now sits in a museum, because the stone turned out to be a 21.25-carat diamond, and Sir Philip Wodehouse, the British governor at the cape, paid van Niekerk's agent the enormous sum of £500 for it. The diamond went to the London firm of Hunt and Roskell, who polished it into a 10.73-carat brilliant, which was christened the Eureka. Still there was no diamond rush.

In March 1869 another Griqua boy found a large crystal, and the next day took it to van Niekerk. It is said that van Niekerk took one look and offered the boy a horse, ten oxen, a wagonload of goods, and a flock of five hundred fat-tailed sheep. The boy accepted this amazing haul, and van Niekerk took the stone. It weighed 83.5 carats and he sold it for £10,000. The diamond became a 47.75-carat oval brilliant, the Star of South Africa, and was sold to the countess of Dudley for £25,000. But before the jewel had even been cut, the colonial secretary, Sir Richard Southey, had it carried into Parliament House in Cape Town. "This diamond, gentlemen," he intoned, "is the rock upon which the future success of South Africa will be built."

As if a gun had fired, the diamond rush began. Seamen deserted their ships in Cape Town and Port Elizabeth. Gold miners arrived from the United States, Canada, and Australia. From Europe came a stream of diamond seekers. Many arrived at Cape Town broke. A straight line from the cape to the diamond fields measured 550 miles, but the route struggled through mountain passes. It took the first fortune hunters months to reach the river junction, and when they got there they transformed the quiet farmland of the Boers into chaos. In less than a year there were fifty thousand diggers working ten thousand claims along the Vaal. The camps had names like Forlorn Hope and Poorman's Kopje, mirroring the wrecked dreams of many. Other diggers got rich: one claim was so stuffed with diamonds that it kept producing for a hundred years.

News of discoveries spread instantly. Men rushed up and down the river like schooling fish. They lived as they could in ragged tents on the veld, freezing at night, or in shacks with carpets spread on the dirt and servants to iron their shirts. When the Standard Bank opened a branch at Klipdrift at the peak of the rush, its safe quickly filled with diamonds and cash. It was a riotous milieu, spouting money, and some of its neighbors thought it needed a firm hand.

The Boers had founded two republics in the vicinity: the Transvaal, north of the Vaal River, and the Orange Free State, east of the diamond field. The Boers were Afrikaans-speaking settlers of Dutch and French descent, religious and upright. They detested the lawlessness of the miners. The Orange Free State was the first to act, asserting legal rights to the diamond rivers. The Transvaal followed suit, declaring that it owned the whole north bank of the Vaal down to the Hartz River. The diggers did not care, as long as they could buy claims and work them. But the Transvaal government challenged them directly, and began to grant to its own citizens mining concessions on land where diggers were established.

The diggers responded at once. They held a series of turbulent meetings that resulted in the proclamation of the Diggers' Republic. As president they elected Stafford Parker, a stern, top-hatted diamond digger who had been a sailor, policeman, and gold miner. Parker was a martinet whose first official act was to appoint a state punisher. The penalty for stealing diamonds was public flogging. Parker ordered all citizens of the republic to take military training. This was wise, for the Transvaal leaders, intent on enforcing their writ, appeared with an armed troop to seize diamond claims. When news of this invasion spread, a large and murderous horde of diggers—the army of the republic—formed and advanced on the Boers.

By now the diggers were seasoned desert rats. They wore the same clothes as the Boers: breeches and wide-brimmed hats.

Many had grown beards to protect their faces from the stinging, wind-blown sand and the brutal sun. They had pistols in their belts. The Boer force retreated. Observing all this, the British at the cape decided to claim the land themselves, and established the colony of Griqualand West, whose chief town was Kimberley. Some diggers wanted to challenge the British, too, but Parker told them they could not fight their own queen, and the Diggers' Republic hauled down its flag for good.

The diamond rush might have merely emptied the rivers of gems and passed into history as a colorful event had it not been for a new idea that began to circulate—that diamonds might be found away from the rivers too. In 1870 a farmer's children in the Orange Free State showed a prospector some stones that glittered in the mud wall of their house on a farm called Dorstfontein; the stones were diamonds. But the farmer hated the riffraff of the camps and refused to let the prospector dig. The prospector went to the next farm, Bultfontein, and soon found diamonds there as well. Then another stone turned up on a farm not far away, Koffiefontein. Hearing about it, the foreman of yet another farm, Jagersfontein, sunk a trench in a dry stream and found a 50-carat stone.

The news exploded into the river camps. Diggers tore down their tents and abandoned the Vaal. The camps were forsaken in a week. Men poured into the Orange Free State. The Boer farmers watched in despair as diggers trampled the vegetation, cut down trees, and stole the cattle. Hotels and bars appeared on the veld. A cloud of dust, visible for miles, hung above the diggings. The owner of Dorstfontein, who had refused even to let prospectors dig, sold up and left.

When word spread that brothers named de Beers were allowing prospecting on their farm, horses and horse-drawn wagons dashed across the veld. The farm was overrun in hours. Every square foot was staked. The brothers finally sold out to a syndicate for £6,300. They had paid £50 for the land eleven years before, and must have felt they had got a good price. They should have asked for a little more, because in the next hundred years the company that made the brothers' name synonymous with diamonds dug £600 million worth of gems from the farm.

The new dry-land deposits produced diamonds in astonishing quantities. South Africa shattered the historical perspective of diamond production. It had taken India two thousand years to produce 20 million carats. Brazil accomplished the same in only two centuries. South Africa did it in fifteen years. That this gusher of jewels did not swamp the diamond price was due to a succession of great magnates whose lives were the stuff of fiction. They fought titanic fights with each other, founded enormous fortunes, and created a modern industry out of a commodity as fanciful as light. The dust clouds had not settled on the plain when the first of these locked horns—Barney Barnato and Cecil Rhodes.

◆

Barney Barnato was born on July 5, 1852, one year to the day before his rival Cecil Rhodes. Only a few hours' journey by train separated their parents' homes, but they came from different worlds. Rhodes was born in the vicarage at Bishop's Stortford, in Hertfordshire. Barnato first saw the light of day in London's East End, in a dilapidated shanty in Cobb's Court at the corner of Petticoat Lane. His father traded in old clothes and scraps of cloth. Barnato grew up sharing a bed with his brother, Harry. He went to the Jews' Free School in Bell Lane, and shortly after his thirteenth birthday, with a bright new penny from his teacher, he left to make his way in the brawling world of thieves and prostitutes and sharp dealers that was his neighborhood.

Barnato took his bold nature from that world, and his name too. Born Barnett Isaacs, he changed his name when he and Harry began to appear as a clown-and-juggling act in the local music halls. Harry's line for introducing Barney was: "And Barney too!" This quickly became Barnato, first as a nickname for Barney, and later as both the brothers' surname.

Barney Barnato. (De Beers)

Barnato was five feet three inches tall, with a stocky build and large, juglike ears. His skin was fair and his cheeks were pink. He had buttery blond hair and blue eyes, and stood out from his darker brother and cousins. He never let anyone slight him. He was quick with his fists, loved women and dirty stories, sold everything from collar studs to elastic bands, played cards for money, and at the age of twenty-one, bursting to do something bigger, he quit his part-time job as a barman at the King of Prussia pub in east London, and headed for Africa.

News of the diamond rush had already drawn one of Barnato's cousins, David Harris, to Kimberley. Then Harry Barnato had gone, and in the late summer of 1873 Barney Barnato arrived at the cape himself. He had £30 in his pocket and a stock-in-trade of forty boxes of bad cigars. He wore a bowler hat and a bright blue jacket and he swung a cane. In that attire he marched out of Cape Town with a Boer farmer who had agreed to take him to

Kimberley for four pounds. He crossed the Great Karoo on foot, and trudged through the mountains. Two months later, his clothes in tatters and his face tanned to the color of dark wood, he swaggered into Kimberley.

Barnato would do anything for money. He unloaded sacks for farmers in the market. Box by box he peddled his "Havana" cigars. When one customer came back to complain that the cigars were the foulest he'd ever smoked, Barnato enrolled him in the enterprise, offering to split the profits if the man would go around Kimberley praising the cigars. Every penny he earned went into trade goods. He bought cloth and combs and penknives—whatever he could turn over quickly. He joined a circus as a prizefighter. When the circus left, Barnato set up a boxing ring of his own and took on anyone for a bet. As soon as he'd scraped together a stake, he began to deal in diamonds.

Every morning at dawn, Barnato headed out of town with his diamond scales and his lens. He plodded for hours through the red mud of the diggings. The dirt caked his skin and his clothes. He thrived on the haggling of the camps, arriving at the diggers' sorting tables with a bottle of Cape Smoke brandy to slake their thirst. He learned to make good trades by making bad, and soon could distinguish a Bultfontein diamond from a lesser gem from the Dutoitspan pit.

The best diamond soil was the prized "yellow ground," a layer of weathered kimberlite oxidized to a tawny yellow color. A maze of small claims honeycombed each deposit. A claim measured only thirty-one feet along each side, and was often subdivided into smaller claims. The Kimberley pipe was a hive of sixteen hundred separate claims, densely cobwebbed with ropeways carrying buckets of soil up from the pits. Footpaths ran out into the digs on fifteen-foot-wide easements. As the diggers went deeper, the high, unbuttressed paths became more treacherous. The thin walls of earth between the claims began to collapse. To prevent this, the government amended the law that had prohibited the consolidation of small claims into larger blocks. Syndi-

Early digs at Kimberley. (De Beers)

cates quickly formed to buy out small diggers and create large claims. This development, and a strange new rumor, prompted Barnato's first big gamble.

He had heard that a local mineralogist was advancing the theory that the diamonds had come to the surface in volcanoes. If this was true, then the yellow ground was only the top layer of the deposit, and more diamonds lay underneath. Early in 1876 Barnato learned that two brothers wanted to sell their claims in the center of the Kimberley excavations. They were running out of yellow ground and did not think the underlying "blue ground" would have diamonds. Barnato bought the claims for £3,000, the entire sum that he and Harry had saved in three years. They brought in diggers and set about penetrating the blue ground. They found only a few stones. Harry became despondent, but Barnato hired more diggers. He said he would dig until it broke him, which it almost did, until suddenly they started finding stones of 10, 15, and 25 carats. In one week they recouped their whole investment and at the end of the year had £90,000.

Barnato bought more claims, and within a few years formed the Barnato Mining Company. The capital was £300,000. A picture of Barnato at the time shows him in a checked suit, with a hand thrust in his pocket and a flower in his buttonhole. He was at the top of his form, and he would need to be, because fate was about to set him on a collision course with one of the hardest heads in the empire, Cecil Rhodes.

Rhodes had delicate health from birth, and the careers he would have liked—the army or the church—were closed. He could not even follow his brothers to the elite boys' schools of Eton and Winchester, but went instead to the local grammar school. At the age of seventeen, with £2,000 borrowed from an aunt, he went to Natal province to learn cotton farming from his brother Herbert. Two months after Rhodes arrived, Herbert caught diamond fever, and Rhodes followed him to Kimberley. Herbert soon tired of the diamond scene and returned to his plantation, but Rhodes remained behind, and bought a claim.

He was an ungainly figure, with pallid skin and long, thin arms that poked from his schoolboy's blazer. He would sit on an upturned bucket while his diggers shoveled away, and lose himself in a copy of Virgil's *Aeneid.* He read Marcus Aurelius. He interrupted his reading only to sort diamonds. In the evening he made them up into packets, and went riding back to town in the gathering night, his rust-red pony picking its way down the road and a tailless dog, his sole companion, trotting alongside.

Rhodes amassed a fortune, mostly by renting out pumps to diggers whose claims were flooded. He used his windfall profits from the pumps to buy claims in the de Beers mine. In 1880 he formed the De Beers Mining Company Ltd.

Rhodes detested Barnato, whom he called "the little prancer." Rhodes was cold and abrupt; Barnato was ebullient. Rhodes had bad lungs and a weak heart. He lived like a monk and slept on an iron bed. Barnato jumped out of bed in the morning and stood outside his shack and swung his Indian clubs. On cold days he broke the ice in his washbasin and poured the freezing water over his head, then shadow-boxed. The two came into conflict because each wanted the same thing—the pipe called the Big Hole.

By 1887 two companies dominated the Kimberley diamond scene—Cecil Rhodes's De Beers, which owned the pit one mile east of Kimberley, and Kimberley Central Diamond Mining

Cecil Rhodes. (De Beers)

Company, which controlled the Big Hole. Barnato had merged his own company into Kimberley Central, becoming the largest shareholder. The next-largest interest in Kimberley Central was held by Compagnie Française des Mines de Diamant du Cap de Bon Espérance, known as the French Company. Barnato wanted to buy out the French Company, but its directors hated him and would not sell. Rhodes knew this and, backed by a syndicate of London financiers that included the Rothschilds, made an offer of £1.4 million to the French. Barnato heard about this, and immediately counteroffered with £1.75 million.

Rhodes saw that this would profit only the French Company. He went to Barnato and suggested that, instead of bidding each other up, they cooperate. He proposed a series of transactions. If Barnato would withdraw his counteroffer, and agree to let Rhodes buy out the French at Rhodes's original bid of £1.4 million, Rhodes would then sell the French Company to Barnato, receiving in exchange a one-fifth interest in Kimberley Central, plus £300,000 in cash. Barnato thought it over. There did not seem to be a drawback. Rhodes would have one fifth of Kimberley Central, but Barnato would have much more. He believed he could easily contain his adversary. So he agreed, and the deal went through. The most implacable man in Africa now had his foot wedged firmly in Barnato's door.

While it was true that Barnato had the biggest share of Kimberley Central, there were small shareholders too, and now Rhodes went after these. He started mopping up whatever shares he could find on the open market. Alerted to this, Barnato too began buying shares. The share price of Kimberley Central soared. In the end it was Barnato who broke. He abandoned the contest, either realizing the depth of Rhodes's financing or content that he had helped drive up the price. He sold his stock to Rhodes. The fight seemed over when, unexpectedly, an obstacle appeared.

A group of Kimberley Central's small shareholders banded together and went to court to oppose the sale to De Beers. They pointed out to the court that Kimberley Central's charter said it could only merge with a similar company. De Beers was not similar, they maintained, because its aims were not simply mining. They cited the De Beers charter, which empowered directors to "take steps for good government of any territory, raise and maintain a standing army, and undertake warlike operations." The court agreed with the shareholders, and ruled that the companies could not merge. Rhodes and Barnato circumvented this by dissolving Kimberley Central. They sold the assets of the company to De Beers and paid off the flabbergasted litigants. The canceled check for £5,338,650, made out to "the liquidators" of Kimberley Central, still hangs in the old De Beers boardroom on Stockdale Street in Kimberley—the check that started the diamond cartel.

Barnato became a life director of De Beers, with a huge block of its stock. He went off to the Witwatersrand gold field, founded a great mining house, and his heirs became millionaire sportsmen and tycoons. Barnato himself was one of the richest men in the

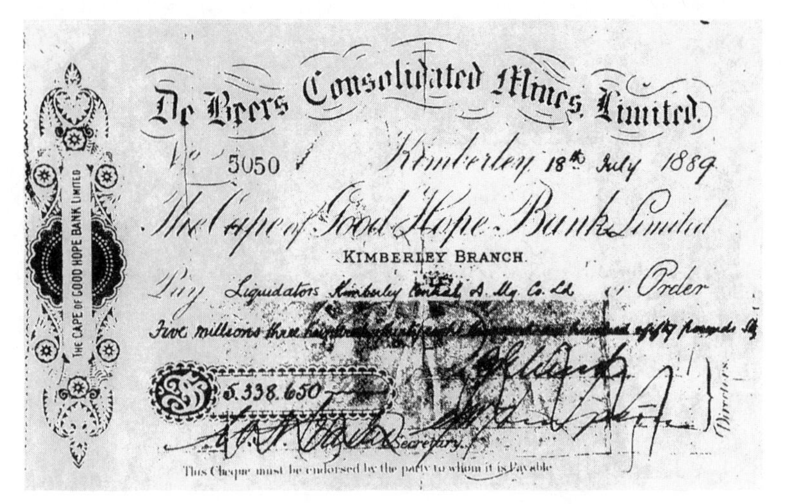
De Beers Consolidated Mines, Limited.
No 5050 Kimberley 18th July 1889
The Cape of Good Hope Bank Limited
KIMBERLEY BRANCH.
Pay Liquidators Kimberley Central D. M. Co. Ld or Order
Five millions three [illegible] fifty pounds [illegible]
£5.338.650
Secretary.
Directors.
This Cheque must be endorsed by the party to whom it is Payable
THE CAPE OF GOOD HOPE BANK LIMITED

The check that started the cartel: With the sudden payment of £5,338,650, Cecil Rhodes dissolved the last opposition to his takeover of the Kimberley mine. (De Beers)

British Empire. He won election to the assembly at Cape Town. Every barrier in life had dissolved before him. But the sheer complexity of his huge business began to tell, and his mental health declined. In June 1897, while sailing back to England on the steamship *Scot*, he either jumped or fell overboard. His death was ruled a suicide. Barnato was a few weeks short of his forty-fifth birthday. Rhodes outlived him by less than five years, but the company he built created the modern diamond business.

◆

With De Beers in control of the great pits, Rhodes did not wait long to show how he meant to proceed. First, he slashed production, and South Africa's diamond output plummeted 40 percent. The price for rough, which had been dropping, turned on a dime and climbed from twenty shillings per carat to thirty shillings. By 1900, De Beers controlled 90 percent of the world's supply of rough. It sold the goods to a syndicate of buyers based in London. The pool of South African diamonds was funneled into London, and De Beers controlled the flow. Only one event could threaten the system: the discovery of new supply.

The vulnerability of De Beers to even a single discovery was plainly shown in 1902 when Thomas Cullinan, a former bricklayer, found the Premier mine not far from Johannesburg. When Sir Alfred Beit, who was allied with De Beers, drove out to see it, he fell down in a faint: The pipe had a surface expression of eighty acres, three times larger than the largest of the Kimberley pipes.

De Beers directors made overtures to Cullinan, which he rebuffed. The company then tried to enlist him in the cartel, warning him that the stability of the diamond price depended on there being a single seller of rough. Cullinan refused that too. He distrusted De Beers. Since only De Beers would actually sell the rough, Cullinan would have no way of knowing whether they were paying him a fair price. Moreover, he had seen the diamonds from his test samples, and shown them to excited buyers in Europe. Cullinan thought he could go it alone. In 1904, its first year of operation, the Premier mine produced 750,000 carats, equal to about a third of De Beers's total from all the Kimberley mines. And there was more to come.

Late in the afternoon of January 26, 1905, as the last shift was coming out of the Premier pit, a miner hurried up to the surface manager, F. G. Wells, and reported that a blaze of light was coming from the pit wall. The rays of the setting sun were obviously hitting something with a crystal face. Wells walked over to the edge of the pit and took a look into the excavation. He saw the light too. The pit had descended to a depth of only thirty feet by then, and the reflection was coming from a point high up on the slope.

The crater walls were steep, but Wells took off his jacket and clambered down to inspect the stone. He had never seen anything like it. When he brushed the dirt away, he saw a clear crystal the size of a fist. Wells pried it out with a pocketknife, scrambled back up the slope, and hurried to the office. The general manager was busy, and while Wells waited, another mine employee asked him what he had. When Wells showed him the crystal, the man gave a derisive laugh, snatched the stone from Wells, and flung it out the window. Wells said nothing, but went outside and retrieved the stone. The man who threw the crystal away was not the only doubter. When a telegram arrived that

Joseph Asscher cleaving the Cullinan diamond in Amsterdam, February 14, 1908. (De Beers)

night announcing the discovery to Cullinan, he told his dinner guests: "I expect they are wrong."

It is not surprising people doubted it, because the Cullinan was the largest diamond ever discovered. It weighed 3,106 carats, or 1.3 pounds. It took three diamond polishers, each working fourteen hours a day, a full eight months to polish the Cullinan into nine jewels. The total weight of the polished gems was 1,055.9 carats, which means that the men sawed off, ground away, and otherwise threw out more than 2,000 carats, or about 65 percent of the weight of the original rough. It was a merciless attack on the diamond, and yielded breathtaking jewels.

Cullinan I, the Great Star of Africa, is a 530.2-carat pear-cut jewel with seventy-four facets, set in the British royal scepter and displayed in the Tower of London. A moving walkway draws visitors past the stone, which from some angles brightens into a sheet of blinding silver light, until the walkway moves on and the angle changes, and the light seems to mass along the hair-thin lines where the facets meet. Then the polished face goes dark, and the eye is left to ponder an unfathomable depth.

The year they discovered the Cullinan, production at the Premier mine increased, as it did again the following year, until it reached 2 million carats annually, roughly equal to the output of

Models of polished gems from the Cullinan rough. Cullinan I, also known as the Great Star of Africa, weighs 530.2 carats—the largest diamond in the world. It is set in the British royal scepter. (De Beers)

all the De Beers mines together. In the first decade of its life, the Premier mine chopped De Beers's share of world production from 90 percent to 40 percent. The only consolation De Beers could take was that it had bought stock in Premier, on the advice of a bold and rising diamond dealer, Ernest Oppenheimer.

Oppenheimer was a representative of the powerful London diamond syndicate that bought and sold De Beers's production. The Premier mine had threatened not only De Beers but the syndicate too, by selling its diamonds to other buyers. One of those fortunate buyers was Bernard Oppenheimer, Ernest's eldest brother, who grew rich on profits that would otherwise have gone to the syndicate. De Beers finally plucked the Premier thorn from its side in 1914, at the outbreak of war. Most mines had closed as Europe divided into warring camps. Premier's share price fell, and De Beers bought control. Once again it ruled the diamond empire. But soon the master of diamonds would face another threat, even more unbiddable than Cullinan—Ernest Oppenheimer himself.

◆

Oppenheimer was a German Jew from a large and well-connected family. His father, a merchant in Friedberg, sent his sons to London to escape a climate of rising anti-Semitism. Oppenheimer arrived in the British capital in 1896 at the age of sixteen, a shy and unassuming boy. He went to work for Anton Dunkelsbuhler, a diamond trader related to Oppenheimer by marriage. Two other Oppenheimer sons, Otto and Louis, already worked at the firm. Dunkelsbuhler was a member of the London diamond syndicate that bought De Beers's production and sold it into the cutting centers of Amsterdam and Antwerp.

"Old Dunkels," as his young staff called him, was short and dictatorial, blind in one eye, with a gleaming bald head and a broad stomach. One day Ernest Oppenheimer, the most junior clerk, was filling inkwells when he stumbled and dumped a load of ink on Dunkelsbuhler's head. Dunkelsbuhler shot to his feet in a rage. "Diamond expert!" he shouted at Oppenheimer. "Why, you wouldn't even make a good waiter!" But Oppenheimer had an aptitude for diamonds. He loved to sort them, and learned quickly. Soon he was training others, and began to sell diamonds too. He rose in the firm, and in 1902 Dunkelsbuhler sent him to South Africa.

The young buyer arrived in Kimberley with excellent introductions. His cousin Fritz Hirschhorn was a De Beers director, as well as an active member of the London syndicate. He also worked for the diamond trading and banking house Wernher, Beit and Company, headed by Sir Alfred Beit. Hirschhorn received Oppenheimer warmly, and at Hirschhorn's house Ernest met the leading figures of the diamond club.

Oppenheimer threw himself into the business life of Kimberley. At least one account of that time details an amazing tangle of self-dealing at the topmost level of De Beers, whose directors were often also its biggest customers. Oppenheimer's connections placed him close to this profitable network. Trading for the syndicate and for himself, he made a large fortune. His blood relations gave him an advantage, and ability did the rest. Perfectly austere in appearance, dignified, and clever, Oppenheimer must

Ernest Oppenheimer. (De Beers)

have seemed to his associates a solid member of their club. Alas for them, the only club that Oppenheimer belonged to was his own.

One of the first friends Oppenheimer made was Solly Joel. Joel was a "Rand Lord," one of the millionaires whose fortunes came, at least in part, from the gold fields of the Witwatersrand. He was Barney Barnato's nephew and business heir, a huge and flamboyant character, a yachtsman and racehorse owner who sported a Vandyke beard. Besides running the family company, Joel was a De Beers director and the most important member of the syndicate. It may have been the example of magnates like Joel that fired Oppenheimer's ambition, for within fifteen years of arriving in South Africa he launched the opening moves of an audacious campaign.

Oppenheimer's first masterstroke was the formation of the Anglo American Corporation of South Africa, and the selection of J. P. Morgan, the New York financier, as his banker. By registering and quartering his new company in South Africa, and allying it with an American financier, Oppenheimer created his own locus of power, independent of such London-based institutions as the Rothschild bank, with its historic links to De Beers.

It has been said of Oppenheimer that he made himself the

world's most powerful mining tycoon by playing his hunches. Perhaps so, but he prepared his way with careful planning and long study and a clear idea of who his quarry was. Although he had made himself, through Anglo American, a power in the gold fields, he never let his attention drift from the diamond business. Diamond money had formed the heart of Rhodes's commercial empire, and Barnato's too, and it would do for Oppenheimer. Controlling diamonds was his goal from the very moment of forming his new company; nothing less would have appealed to his predatory nature. Control of diamonds meant control of De Beers. The long stalk began.

Oppenheimer opened several fronts, among them the careful husbanding of his relations with Solly Joel. Joel was not only the largest shareholder in De Beers but the leading member of the diamond syndicate. Oppenheimer knew he could not have every hand against him as he set about conquering the diamond world, and Joel would be an invaluable ally.

At the outbreak of World War I the diamond business was in a bad state. Mines closed, the price plunged, and the syndicate was left with a stockpile it could not afford to sell. In 1915 South Africa seized the diamond-rich German colony of South West Africa, present-day Namibia. The colony became a South African protectorate. Although diamond production in South Africa itself had shut down, the administrators of the protectorate, wanting any revenue they could get, kept the diamond mines running. By 1919, at the end of the war, the protectorate's diamond production had increased to 18 percent of world supply. So important had these mines become that the South African administrators felt the operations should be amalgamated into a single enterprise. Under pressure, the German owners agreed to sell.

Friends of Oppenheimer, no doubt with his encouragement, approached the South African prime minister, General Louis Botha, and suggested Oppenheimer as a buyer. Botha agreed to consider it. Then Fritz Hirschhorn, as a De Beers director—and knowing nothing of Oppenheimer's move—paid his own visit to the general. Botha told him that "certain mining interests" had approached the government already. Hirschhorn immediately realized what was up and sent an urgent cable to De Beers's London directors. Amazingly, the London directors failed to recognize the threat that Oppenheimer posed. They blithely cabled back: "German holders under belief they will remain in possession of their property and from our inquiries in Germany these holders are not disposed to sell at present. We think under these circumstances [Oppenheimer's] mission will be a failure."

Their belief cost them Namibia. Oppenheimer bought every diamond mine in the protectorate, a gigantic diamond lode. His new company would develop the Diamond Coast into the greatest treasure chest in Africa, as it remained for decades. It was a brilliant coup. Hirschhorn was outraged. He would soon be even angrier, for Oppenheimer and his brother Louis completed a contract with the company they set up to own the mines, which gave the brothers the right to establish a selling syndicate of their own. Put another way, Oppenheimer had agreed to sell his diamonds to Oppenheimer. Hirschhorn was a leader of the existing syndicate, as well as a De Beers director and Oppenheimer's relative. It was as if Oppenheimer had publicly shredded his past life and thrown the pieces in his old friend's face, a rejection that became even more complete some years in the future, when Oppenheimer left the Jewish faith and converted to the Church of England.

Five years after his Namibian coup, Oppenheimer, now Sir Ernest, readied for the final assault on De Beers. He had already spelled out his intentions in a letter to J. P. Morgan: "From the very start I expressed the hope that besides gold we might create step by step a leading position in the diamond world, thus concentrating by degrees in the Corporation's hands the position which the pioneers of the diamond industry (Cecil Rhodes, Wernher, Beit and Company, etc.) formerly occupied." He began to increase his holdings in De Beers.

Oppenheimer steadily added diamond assets to his portfolio. He had kept his relations with Solly Joel in excellent repair, neutralizing a potential enemy. Together the two contracted to buy the entire diamond output of the Belgian Congo. Oppenheimer took an interest in various fields in West Africa, along with the right to market the production. He joined the syndicate, but it promptly expelled him when he bought for his own account a huge parcel of goods that the syndicate thought it should have had for itself. Oppenheimer replied by forming, yet again, his own syndicate. He brought under his control huge new alluvial diamond fields in Namaqualand, on the Atlantic coast. He bought still more De Beers stock. His stature in the diamond world was unmatched.

The chairmanship of De Beers, the prize he sought, should have fallen into Oppenheimer's hands like ripe fruit. Although De Beers was leaderless, those who opposed Oppenheimer's accession to the throne of diamonds, including the Rothschilds, feared the concentration of so much power in one man's hands—chairman of De Beers and head of the new syndicate, *his* syndicate. Finally, through the sheer weight of his position, Oppenheimer prevailed. On Friday, December 20, 1929, the De Beers board unanimously elected him to the chair.

In the old boardroom on Stockdale Street in Kimberley, Oppenheimer took the high seat of diamonds. He was forty-nine years old, at the peak of his power. Pictures of Barney Barnato and Cecil Rhodes gazed down from the wall. A director made a curt, formal speech. Sir David Harris, Barnato's cousin, sat grimly in his chair. Fritz Hirschhorn yanked at his mustache and glared at the table. Apparently Oppenheimer was perfectly untroubled by the animosity, and sat there with his right knee grasped in his hands in a boyish pose of satisfaction.

◆

Sir Ernest Oppenheimer destroyed the London diamond syndicate. From the moment he gathered the reins of the diamond empire into his hands, the doom of his old friends was writ with an iron pen. He knew them too well. Never again would De Beers sell its rough to middlemen: it would sell the goods itself.

De Beers called the system Oppenheimer started "single-channel marketing." A cartel of producers sent its rough to London to be sold. The producers were either De Beers mines, wholly or jointly owned, or mines that contracted with De Beers to sell their goods to the cartel. In addition, the company mopped up loose rough wherever it came to market. De Beers mixed it all together in London, sorted it, prepared a number of selling mixtures known as "boxes," and sold it to clients who paid what they were told. At its peak this system controlled 80 percent of the world's rough. If the diamond price weakened, De Beers cut back the flow of goods into the cutting centers; when the price recovered, it opened the tap again.

The magnitude of the diamond empire gave weight to a tacit creed—that only De Beers knew what was best for diamonds. In light of the sheer massiveness of the cartel's position in the rough market (one could say it *was* the rough market), it seemed reasonable to infer that damage to De Beers was damage to diamonds. This neat turn of logic helped De Beers survive what might have been a crippling challenge when, in 1954, Soviet geologists uncovered a cluster of pipes on the Siberian Craton. The Soviets were intensely secretive about the discovery, and there was speculation that they would not be able to recover the diamonds profitably. After all, the ground was permanently frozen to a depth of 350 feet and no transportation or power infrastructure served the site. But any thought that these considerations would dissuade the Soviet Union was a barren hope. It had no investors to appease, and did not care what it cost to get the diamonds. What it cared about was the foreign cash the diamonds would bring in.

The Soviets built a mining city, stripped the forests from the top of the pipes, and started digging. De Beers executives flew to Moscow to persuade the communist leadership to join the cartel. The company had reason to fear the Siberian pipes; they were fantastically rich and would come to supply a quarter of the world's rough market. De Beers warned Moscow that if this production were dumped into Antwerp and Tel Aviv, the diamond price would crumble. Moscow concurred, and signed a marketing agreement with the cartel.

The Soviets' only demurral about dealing with De Beers came after the events of March 21, 1960, when South African police fired on a crowd of blacks in the township of Sharpeville, killing sixty-nine. In the storm of outrage that followed, Moscow felt embarrassed by its link to a South African company controlled by a white billionaire resident in the racist state. A new company, City and West East Ltd., was immediately formed to buy the Siberian diamonds. It had no visible ties to De Beers, but they owned it. The diamonds flowed to London as before.

The empire of diamonds—the great middle empire between the past and the present—had as its whip hand the control of rough. But the whip could be snatched away. De Beers had succeeded for a time in guiding the Russian goods into London, but the relationship with the Russians was uneasy. Moscow's diamond czars distrusted De Beers's valuations of the Siberian production, and as Russian diamond competence grew, so did the chance that the Russians might elect to market the rough themselves. De Beers's best protection against such a development was to have more diamonds than anyone else, which empowered the cartel to punish rebellious producers by flooding the market and depressing the price. With its great size and diamond hoard, De Beers could weather such choppy waters better than its foes.

A central mission of De Beers, then, was to locate new mines before others could locate them. The company sent its geologists into many lonely parts of the world, where they might remain for years. The diamond empire rested on manipulation, certainly, but also on long searches in empty places, and on diamond knowledge pure and simple.

◆

In 1955 Gavin Lamont, a blade-thin South African with a lean smile and a quiet manner, began a long diamond search in the Bechuanaland Protectorate, now Botswana. Pictures from the time show the veteran De Beers geologist dressed in a short-sleeved khaki safari jacket and khaki shorts. A scarf is knotted rakishly at his throat. His white hair is combed back neatly and he seems always to be holding his hat at his side while he squints in the sun.

Lamont set up his headquarters in the town of Lobatse, on the southern edge of the main geographical feature of Botswana, the Kalahari Desert. The Kaapvaal Craton, which hosts South Africa's diamond pipes, extends beneath the Kalahari. If the cra-

ton hosted diamond pipes a few hours' drive to the south, reasoned Lamont, why not here?

The desert covers 200,000 square miles. In some places the sand is one hundred feet deep. If there were pipes in this part of the craton, the evidence for them would seem to be irretrievably smothered by that massive quilt of overburden. Every diamond indicator mineral would be packed away under yards of sand, and the task of digging through the Kalahari in search of them seemed a hopeless one. But Lamont thought he saw a way.

Heaped around on the Kalahari were tall, red anthills. Colonies of ants lived in the desert, as they had for thousands of years. They built networks of tunnels that extended far below the surface. Over successive generations, billions and billions of ants tunneled away, digging out grains of sand and other minerals and carrying them to the surface to be added to the rising hills. The ants were sampling the deep Kalahari. In time, each anthill would be leveled by weather, but the minerals ferried up from below would remain in the topsoil of the desert. The surface layer was therefore impregnated with minerals from a great depth, and to sample the surface would be to sample deeper soils too. If there were a pipe below, there ought to be diamond indicators on the top.

(Lamont's exploration took place long before John Gurney had discovered the relationship of garnet chemistry to diamond prospectivity, but even in those earlier days geologists knew that garnets were among the mineral grains that sometimes pointed the way to diamonds.)

Lamont deployed samplers into the desert. They cut down thorn trees until they had a straight line they could drive their truck along. Then they paced out a grid and sampled the desert at set points. Lamont began receiving reports of indicators. The grains were widely dispersed, and no samples turned up high concentrations, but to Lamont the presence of the indicators meant that a pipe existed somewhere beneath the Kalahari.

He searched the desert for six years. Samplers walked the lines of their grids, pushing along a bicycle wheel with an odometer attached to measure off the distance between sampling points. The results were inconclusive, and in 1961 Lamont called his teams out of the Kalahari and turned his attentions to the east.

Two years previously, at the edge of the Kalahari, another company had found three diamonds. It had failed to locate a source, and had eventually abandoned its prospecting concession. Lamont went to the site and quickly found diamonds and indicator minerals. The deposit suggested a river origin for the diamonds, and Lamont's samplers combed the terrain for any sign of where the diamonds came from. They found nothing. Lamont went over the area again and again but, like those before him, could not find a source. He wrote off the diamonds as an anomaly, and moved his samplers elsewhere.

Also based in Lobatse at the time was a government geologist, Chris Jennings. Jennings remembers long conversations with Lamont, whiling away the evening with speculations about the possibility of pipes in that part of the craton. They returned often to the subject of the diamond occurrence at the edge of the desert. Then one day Jennings recalled a paper he had read at university, in which the writer had postulated a series of uplifts across southern Africa—warps in the surface of the Earth. Could such an uplift have cut off the flow of the diamond river, concealing the location of the diamonds' source? He mentioned the paper to Lamont.

In 1965 Lamont returned to the site. The diamonds had been found just east of the eastern edge of the Kalahari. A low escarpment marked the end of the desert. If the uplift theory was right, an ancient river flowing southeast out of the Kalahari, and draining it, had suddenly been cut off by a lift in the Earth's crust. The lift was marked by the escarpment. If this hypothesis held, then the upstream source of the diamonds lay back in the Kalahari, buried under sand. Lamont ordered a series of traverses through the Kalahari scrubland west of the escarpment. In July of 1966, in the Orapa area west of the village of Letlhakane, the geologist Manfred Marks found a dense concentration of indicator minerals and a large, round, surface anomaly. They sank a pit in the sand and found the gray-green soil of a kimberlite.

The top of the Orapa diamond pipe covers 278 acres, about the area of twenty-five blocks of midtown Manhattan. It produces 12 million carats a year. The find led to discoveries at Letlhakane and Jwaneng, and to present-day Botswana's position as the top diamond-producing country in the world, with 25 million carats of rough a year worth some $3 billion.

Botswana's gusher of rough is the greatest diamond source of the De Beers cartel. As the diamonds flow to London they are joined by volumes from the company's other wholly owned or jointly owned mines. In 1999 De Beers took 1.3 million carats out of Namibia and 230,000 carats from Tanzania. The South African mines added their own streams to the river: 3.7 million carats from Venetia, 1.7 million carats from Finsch, 1.5 million carats from Premier, 736,000 carats from Kleinsee, and 563,000 carats from Kimberley. Altogether, adding the rough from the smaller mines into the mix, southern Africa gave the cartel more than 34 million carats of diamonds in that single year. Worth about $4 billion, it formed the platform from which the diamond empire ruled its world.

4 The Long Hunt

At the time of the Orapa discovery, with the greatest diamond lode in history soon to come flowing into the cartel, a long and remarkable chain of events began. It had thunderous consequences for the cartel. When the implications finally became clear, the borders of the great diamond empire would be thoroughly breached, its ability to control the diamond world in tatters, and a new corps of diamond hunters spoiling to contest it in the field. It would have been astonishing to anyone at the time of Orapa, with the fortunes of the empire so bright, that events destined to reduce it could begin with the actions of a government geologist in Lobatse.

In those days Chris Jennings could often be found bumping through the desert in a Bedford five-ton truck. Beside him in the hot cab sat his wife, Jeanne. The rainy season brought torrential downpours to the desert. The soil would turn to mud. Some days it took eight hours to move a quarter mile. Finally the Bedford would bury itself to the axles. When they could go no farther they made camp, and Jennings, tall and athletic, would go slogging off into the Kalahari, making notes.

One of Jennings's duties was to map potential water sources. When rain fell in the desert, water seeped down and collected in deep basins. The challenge was to locate them, but Jennings had none of the sophisticated geophysical instruments that would have made this easy. Instead, he ordered the carpenter in Lobatse to construct a pair of wooden boxes. Into each box he positioned twenty-five batteries, of forty-five volts apiece. Each box could therefore deliver an electric "kick" of more than a thousand volts. They loaded the boxes into the Bedford, and when they reached a likely depression they unloaded the boxes, sank electrodes into the sand, and sent down a current. By trial and error, Jennings figured out the time it took for the current to bounce back from a potential basin. The deeper it was, the likelier it would hold water. Jennings's weird contraption was well known in the small geological community of Lobatse, and when Gavin Lamont discovered Orapa, he invited Jennings to lug his unwieldy appliance onto the site to see if the pipe had a geophysical reflection.

Jennings piled his batteries into the truck and drove out to Orapa. He camped on the pipe for two weeks, tapping steel leads into the sand and sending down surges of current. He discovered that the pipe did produce results that distinguished it from the surrounding desert and constituted a distinct geophysical signature. The implication struck Jennings: find a signature like Orapa's, and there would be the next Orapa. A few months later, after fourteen years in Lobatse, the Jenningses left Botswana. By then they had a family, and the children would need higher schooling. Jennings exchanged the footloose life of field geology for the daily grind of Johannesburg.

But he could not stop thinking of the pipe. He had moved to the city to become head of African exploration for Falconbridge Limited, a Canadian multinational mining company. His targets

Chris Jennings placing a geophysical probe in Botswana in 1965. (Jennings family)

were nickel, copper, and tin, but the South African geologist does not exist without diamonds burned into his brain, and Jennings's thoughts often drifted back to Orapa. At Falconbridge, he had resources far beyond those of a civil servant in a remote protectorate. As he read more and more about developments elsewhere in the world, his ideas about a diamond hunt in the Kalahari began to coalesce. In Canada, for example, Falconbridge geologists were using airborne geophysics to search for base metals. The data that Jennings had retrieved with his truckload of batteries was being collected by pods of instruments towed behind aircraft. Jennings realized that the airborne technique, allied with what he already knew about the geophysical reflection of pipes, would enable him to examine enormous tracts of the Kalahari far more rapidly and confidently than Gavin Lamont had been able to do.

It is easy to imagine Jennings's restlessness in Johannesburg. In an office he projects suppressed energy, his large hands always darting here and there to pick up a pencil and tap it on the desk, or to open and shut a drawer. He jumps to his feet to consult maps posted on the wall. He seems anxious to be gone, into some place where long strides will release him from unbearable confinement.

Jennings had collected Orapa's signature while still a civil servant, or Lamont would not have allowed him near the pipe. For ethical reasons, then, Jennings waited four years before he put his knowledge into action. Finally, in September of 1974, he made his pitch at a special meeting in Johannesburg attended by the chief executives of Falconbridge and of Superior Oil, a Texas company that owned a large piece of Falconbridge. For two hours he poured out everything he knew about diamond mines, including the potential profits. He painted a picture of an industry with an insatiable appetite, explaining that the appetite was insatiable for the plain reason that De Beers would not allow it to be sated. The company ensured that supply fell short of demand. Any supplier entering this market outside the De Beers channel would find an eager market.

Then he unveiled his plan. He told his listeners what he had learned about Orapa, and how swiftly a campaign of airborne geophysics would reveal hidden pipes. He stressed that pipes occurred in clusters: if they found one, they would find others. The visitors were convinced. On the spot, Superior Oil's chief proposed a fifty-fifty joint venture of Falconbridge and Superior Oil. The diamond chase was on.

The air campaign over the Kalahari commenced a few months later, when a DC-3 loaded with geophysical equipment taxied away from a private hangar near Johannesburg, lumbered down the runway, and lifted into the dawn sky. It headed north for Botswana. For the next thirty months Jennings ran a steady schedule of flights over the Kalahari. The plane flew a grid pattern. At Falconbridge they kept a tight lid of secrecy on the exploration, but De Beers found out. The regular flights of the old DC-3, appearing low above the Kalahari and droning along in a straight line until it disappeared, would have been hard to miss. Soon enough it would return, still low on the desert, tracing a return path parallel to its former course. De Beers field staff reported the flights to Johannesburg. Jennings heard from a source at De Beers that the Falconbridge flights had been discussed at a directors' meeting, and had raised the ire of the company's top executives. As Jennings put it later: "The continent of Africa was their property. You never had to wonder if you were in their backyard because everything was their backyard." The feeling in the De Beers boardroom might have turned from irritation to dismay if they had known what Jennings had turned up—targets by the dozen.

In all, Jennings's team found sixty-six pipes. The explorers could scarcely believe the number. At one point they were finding one pipe every week. The mood at Falconbridge's Johannesburg headquarters was ecstatic. But as pipe after pipe showed up, and the tally grew, Jennings realized that the size of the success, by itself, presented a problem—cost.

Few pipes contain high enough concentrations of diamonds (grade) to support a mine. Of more than six thousand known pipes in the world, most are barren. The odds against a kimberlite pipe containing economic quantities of diamonds are more than 200 to 1. The exploration of a pipe proceeds by extracting samples of kimberlite, which are processed for diamonds. Since the gems are distributed unevenly and randomly throughout the host (a characteristic called "nugget effect"), a satisfactory diamond count from an early sample will lead to the extraction of progressively larger samples, until the miner is convinced that the sample he has taken is large enough to support a reliable extrapolation of a value for the whole pipe.

In the Kalahari, a pipe would be covered by an overburden of sand between fifty and two hundred yards deep. Jennings calculated that the removal of overburden would push the bill for the sampling of a single pipe to $10 million. To pick its way through every pipe, the joint venture would face a ruinous bill. Jennings needed some way to decide which pipes had the best chance of containing diamonds. He called John Gurney.

Gurney faced a situation new to diamond exploration: not a single pipe, but sixty-six of them, to contemplate. The challenge was to isolate the best targets by collecting indicator minerals from each and analyzing the grains. But while Gurney was confident in his conclusions about the association of the G10 with a

John Gurney. (John Gurney)

diamond-bearing pipe, he did not think that a predictive comfort level had been reached. He needed more data, from more pipes. This requirement launched an exhaustive survey of diamond pipes. Gurney and Jennings assembled every shred of information on every pipe they could. Where access to pipes was denied, Gurney combed the scientific literature for clues. He erected a substantial platform of data, but there was one glaring weakness: He had no minerals from the only places in the Kalahari where they *knew* there were diamonds—the De Beers pipes at Orapa, Letlhakane, and Jwaneng. To obtain examples of the indicators at those sites would be to gain a powerful lens into the Kalahari. When an opportunity to get such indicators presented itself to Jennings, he seized it.

In Botswana on a business trip, Jennings flew over Jwaneng in a small plane. De Beers was still sampling the pipe. Piles of kimberlite lay all over the Kalahari. Jennings made a mental note of where the roads were, and which roads came closest to the piles. When he got back to Johannesburg he tossed a few tins of food and a groundsheet into his Chevrolet pickup, and next

Extracting an ore sample from an African kimberlite target.
(Matthew Hart)

morning at first light he headed north, beginning the adventure that his wife and children still refer to as the midnight raid.

He crossed the border and made his way to Lobatse. It was noon. He drove cautiously through town, taking side streets to avoid being seen by anyone he knew. From Lobatse he took the road to Ganzi. Six hours later he reached the turn that led out into the Kalahari scrub toward Jwaneng.

"The road was west and the sun was in my eyes. Half an hour later, I saw a truck approaching through the desert. Even with the sun in my eyes I knew it was a De Beers truck—one of those big Ford F150s. I thought, God, what if it's Gavin Lamont? So I pulled my hat low and sort of scrunched down in the seat, and just waved vaguely as they went by."

Jennings thought he knew where he was going, but by nightfall he was lost in a maze of rutted tracks that wound through the thorn trees. He drove down one track, and then another, searching for some indication of where the discovery lay. The desert night came down like a black curtain. He was almost ready to give up when a local man appeared in the headlights and gave him a friendly wave. Jennings asked him if he knew the way to the Jwaneng digs, and the stranger agreed to show him the way. They drove the last few kilometers in the dark. Jennings spread his groundsheet in thick bushes and set his clock for midnight.

When the alarm roused him, Jennings crept out into the scrub and made his way to the perimeter of the Jwaneng property. A high fence surrounded the site. Some of the piles of kimberlite were just inside the fence, and material had spilled through the wire and was lying outside. Jennings had a mineral scoop and some sample bags. In an hour he filled four bags, each with fifteen pounds of kimberlite. He left at sunup. By nightfall he was coming down the N1 expressway into Johannesburg, and the next day the sample went to Gurney.

In his lab at the University of Cape Town, Gurney collected all the data and prepared his report. In it, he analyzed forty-nine different kimberlite occurrences. There were more than twenty-five hundred mineral analyses done on the electron microprobe, 230 graphs and plots, and numerous maps and tables. Gurney had also evolved mechanisms for scoring the relative importance of garnet, chromite, and ilmenite samples. Altogether, the document constituted the most powerful diagnostic tool then in existence for predicting diamond grade in target pipes, and it is widely held that the formulas spelled out by Gurney in the secret paper were superior to any then in use by De Beers.

In 1981, seven years after Jennings had sent the first flight into the Kalahari, Falconbridge completed its assessment of the Kalahari targets and picked out the best—a pipe between Jwaneng and Orapa, at a place called Gope. Gurney had found that most of the Kalahari pipes were either barren or low-grade, but when he saw the indicators from the Gope pipe, he knew they had found a mine-grade deposit: "I called it the minute I saw it. I put stars all over it."

This should have been the moment of triumph for the venture: they had combed the desert by air and on the ground, and produced a discovery. But suddenly it flew from Jennings's grasp. The chief executives of Falconbridge and Superior Oil appeared in Johannesburg and began to meet with Harry Oppenheimer, Sir Ernest's son, the De Beers chairman. Jennings was excluded from the meetings. His superiors ordered him to remain on standby in a nearby hotel room to answer calls. "They didn't consult me at all," he said. "Every now and then I'd get a call, they'd ask me some question, and then they'd go back to their meeting with Oppenheimer. What I didn't know was that Oppenheimer was telling them what a risky game diamonds were, that the price was very fragile, the profit margins low, and if the market lost faith in the ability of De Beers to run it, the whole thing would collapse. But I think the most important fact was Oppenheimer himself, a famous billionaire, the emperor of diamonds, and these guys were just totally sucked into a desire to do a deal with him and be part of the cartel."

Falconbridge and Superior Oil gave up a one-third interest in the Gope pipe to De Beers in return for De Beers managing the project. The best new target in the Kalahari was consigned in an afternoon to the people with the least interest in pushing it to development. There could be no advantage to De Beers in bringing to production a mine whose profits would only encourage Falconbridge and Superior to look for more deposits. Said Gurney: "The rug was snapped from under everyone's feet. The day De Beers took over was the day my income stopped." Gurney put away his data on the Kalahari and looked for other work. Jennings transferred to Falconbridge's head office in Toronto, where his duties once again were to search for base metals. He had a sour taste in his mouth. But while it seemed as if De Beers had swept its challengers away for good, it had not. Instead, the challenge had moved to another continent.

◆

In 1978, three years before the joint venture surrendered its pipe to De Beers, Superior Oil had opened a separate diamond exploration in North America. The program had never taken on the sound and fury of the Botswana campaign, but had proceeded steadily. North America is strewn with diamonds. They occur in glacial moraines around the Great Lakes; stones have been found in Michigan, Wisconsin, and Kentucky; a diamond mine operated briefly in Arkansas; kimberlite discoveries at Kirkland Lake in northern Ontario attracted attention in the late 1960s, but no mine-grade deposit had been found. North America retained its reputation as a good place to sell diamonds and a bad place to find them.

Hugo Dummett, a senior geologist at Superior Oil, thought differently. Three large cratons underlay the continental crust of North America, and pipes had already been discovered. Dummet thought these pipes should be reassessed according to Gurney's system. He targeted a pipe in Colorado where the United States Geological Survey had found a diamond. He secured the exploration rights, built a diamond-recovery plant in nearby Fort Collins, and sent in a Canadian geologist named Charles Fipke to sample the terrain.

As in South Africa, a regime of secrecy ruled the enterprise. Dummett's position at Superior had entitled him to a copy of Gurney's secret report. Dummett had a lively sense of the value of Gurney's discoveries, and not even Fipke, hired for his skill as a sampler, was told that Dummett was looking for a precise high-chrome, low-calcium signature in the garnets he collected.

Fipke caught diamond fever anyway. Since he knew the target mineral was diamond, it was natural to wonder how the minerals he collected fit into the search. Prospectors are professionally sly. They contend not only with hypotheses, which are often wrong, but also with competitors who try to discover what they are up to, just as they themselves, in turn, spy on those same competitors. As an experienced prospector hired to search for specific minerals by people chasing diamonds, Fipke would naturally wonder whether he himself, simply by looking for these same minerals, might find diamonds of his own. So that is what he did.

In 1979 Fipke was sampling near the town of Golden in the

Rocky Mountains of British Columbia. It was summer, and Fipke's wife and children had come along. One evening while the children slept, Fipke's wife, Marlene, found a chrome diopside in a stream. It was bright green and as large as a child's marble. Fipke, recognizing the mineral as a diamond indicator, told Dummett about it. Dummett agreed to pay for the exploration in exchange for an interest in any discovery. Fipke then telephoned Stuart Blusson, a Vancouver geologist, who soon came racketing into Golden in his old Hughes helicopter.

Blusson had been working for the Geological Survey of Canada for twenty years when he arrived in Golden to help Fipke look for the source of the chrome diopside. He had mapped the geology of large tracts of the Canadian west, and had an intimate familiarity with the stratigraphy of the Rocky Mountains. The day after he arrived, as the first sunlight streamed over the peaks, they flew out of Golden to look for the pipe.

"Marlene had got the diopside from a stream," said Blusson, "so we flew up to the head of the stream. There was a glacier up there, and that's where we found the first pipe, right at the toe of the glacier. I said, 'Hey, Chuck, there's one right there.' It was round, and so small you could throw a rock across it, a dark circle of nice, fresh, polished rock where the glacier had just receded."

Flying up the valley, the prospectors were framed on either side by cliffs of layered rock. Blusson examined the neat formation for signs that it had been disturbed. He found one right away—a place where the even stratigraphy had been broken by some subsequent intrusion. "I knew it was where a pipe had come through. It was there on the cliff, and we [were] looking at the top of it. I could actually make out the shape. It was like looking at the ghost of a pipe."

The erupting kimberlite had smashed through the rocks. Broken rock had then collapsed into the crater, and this patch of rubble in the neat pattern of the cliff betrayed the intrusion. On that first flight they found seventeen pipes. By the time they were through with the area around Golden, they had found twenty-six. In the end the pipes were barren, but neither man could forget the wild exhilaration of that first flight up the valley, with pipe after pipe clearly standing out against the orange cliffs. Blusson quit his government job and joined the diamond hunt.

In 1982 Fipke, Blusson, and Dummett were at a remote camp in the northern Rocky Mountains, examining a kimberlite pipe. Dummett had to fly out on business, and on the way he happened to ask the pilot if the man was flying for any other exploration projects in the area. The pilot said yes, that an outfit called Monopros had hired him. Dummett became very alert. "I said, as casually as I could, 'Where are they?' and he said, 'Oh, they're over on the east side of the Mackenzie River, in the bush.'" Dummett said nothing more. When they landed at Norman Wells on the Mackenzie, he bought a map, marked it, and sent it back to Fipke and Blusson with instructions to go and take a look. For Dummett knew, as the pilot had not, that Monopros was a wholly owned exploration company whose orders came from Johannesburg: Monopros was De Beers.

Fipke and Blusson were astounded. No one had heard a whisper that De Beers was in the region. They learned that De Beers was in the fourth year of an exploration that had carried the company's geologists steadily northward along the east bank of the Mackenzie. Fipke and Blusson discovered that some thirty samplers and field staff—a large camp—were under canvas at Blackwater Lake. Something very tempting must have brought them there in such numbers. Fipke summoned a helicopter, and he and Blusson flew east across the Mackenzie River. In the distance rose the mountains of the McConnell Range. Between the river and the mountains, in thin bush, lay Blackwater Lake. As they approached the De Beers camp, they saw the sampling lines cut through the trees. In a clearing by the lake stood a row of white tents. They flew directly over the camp, and a figure rushed from a tent and trained binoculars on them. They flew away.

At two o'clock the next morning, still daylight at that latitude in summer, they left their camp, again by helicopter, and flew across the river. They landed out of sight, to the east of the

De Beers encampment. Streams ran through the property from east to west, draining the foothills of the McConnell Range into the Mackenzie. The prospectors had placed themselves upstream from the camp. They collected soil from the streams flowing into the concession. Whatever De Beers was collecting, they would collect it too.

When Blusson looked through the samples, he found material much older than the local rocks. He thought the source of the older material must be the Slave Craton, east of the mountains, and that the rocks had been carried into the valley in a glacier. Blusson wanted to explore "up ice," that is, against the direction of glacial flow. By tracing back up the glacier's route, he believed, they would find the source of the rocks and indicator minerals that the glacier had captured. In the meantime, they collected three 20-pound bags of soil and shipped them to Fort Collins. When Dummett's lab examined the samples, they found G10 garnets with Gurney's classic diamond-friendly composition. They found eclogitic garnets, too, which meant the source was also rich in high-grade eclogite. Dummett took the results to the chief executive of Superior Oil and asked for permission to expand his activities. But instead of encouragement, he was told to abandon the search.

It was late in 1982. The joint venture founded in Johannesburg had run for eight years and spent many millions of dollars. The companies' chief executives were frankly tired of the hunt. They had done their deal with Harry Oppenheimer the year before, and the idea of paying for people to scrabble around in the north of Canada struck them as absurd. Chris Jennings, who had been following from Toronto Dummett's diamond progress, was bluntly ordered to keep his mind on base metals. In Houston, Dummett got a similar message: drop the diamond hunt. Jennings and Dummett were disconsolate. They believed the diamond prospects were high, a belief supported by the presence of De Beers. Dummett and Jennings would have been even more enthusiastic had they known that De Beers had already been raking through Canada for thirty years. But there was no reprieve.

Dummett's last act as head of the North American diamond search was one of impulsive generosity: with Superior Oil's consent he gave Fipke all the field data. He held only one thing back—Gurney's report, which held the secret of the G10 and the G9, another diamond garnet. Neither Fipke nor Blusson knew about these two key garnet types. Instead, Dummett invited them to send their samples to Houston for free analysis. The prospectors turned their search eastward.

In moving to the east of the mountains the partners confronted a great expanse of water and rock. The rock was hatched all over with gouges left by glaciers. High ridges of gravel deposited by meltwater wound across the granite in graceful serpentines. The whole land was a script written by vanished ice, which the two must decipher.

The Laurentide ice sheet formed in Keewatin, the region adjacent to the west coast of Hudson Bay. The temperature dropped and the snows fell. The ice cap rose above the land and began to creep to the west. Sometimes it slithered slightly northwards, sometimes slightly south, but kept to its primary westerly course until it lay in a solid sheet more than a mile high that reached to the Mackenzie River. For almost one hundred thousand years the glacier lay on the land, until, some ten thousand years ago, heat from the Earth's core began to melt it. The glacier retreated, and what is visible today in much of Canada is the scrubbed granite surface that it left behind. The rock is Precambrian in age, and the structure is called the Canadian Shield. A hundred miles north of Toronto it tips up into the light of day, revealing the true nature of the country, and if geography can give a place its soul, then there is Canada's, a vast rock plain that rises out of the farms and forests in the south of the country and surges away to the north, shedding the deciduous trees from its forest, then thinning out the pines, then scraping the trees off altogether until at last, perfect in its emptiness, it vanishes into the Arctic sea. The land up there is called the Barrens.

Europeans arrived in the vicinity in the autumn of 1670, when a pair of English ships came tacking into the waters of

Hudson Bay. The colors of the Hudson's Bay Company fluttered at the masts. The crews put ashore at the mouth of the Nelson River and nailed the king of England's coat of arms to a tree. A royal charter gave the company the right to trade in and govern all the lands in the watershed of Hudson Bay. They built forts and set about gathering furs and enduring the appalling climate. Hudson Bay was choked with ice for most of the year. Its waters were not moderated by any ocean current, and winter temperatures fell to minus eighty degrees Fahrenheit. In the decades that followed their arrival, the English huddled by the shore and made few forays inland.

While European exploration had already penetrated other parts of Canada, the first explorer did not head west from Prince of Wales Fort on Hudson Bay until 1769. On November 6 of that year, before dawn, Samuel Hearne set out in a falling snow. Hearne and his small company fell in with a Chipewyan Indian known as Captain Chawchinahaw. Three weeks later, the explorers and Indians had reached a point two hundred miles north of the fort, when suddenly, without warning, Chawchinahaw and his band seized all the food supplies and walked out of camp, "making the woods ring with their laughter." Hearne and the others survived the march back to the fort by eating their buckskin jackets.

Hearne picked the month of February for his next trip out of the fort. The party had only a moosehide tent, a quadrant, some guns, and trading trinkets. They struck north for Baker Lake. The cold was so fierce that the act of breathing was an agony, like drawing ice into their lungs. Even the forest animals seemed overwhelmed by the cold: Hearne captured marten by falling on them. The Englishmen survived the winter, and in June came out on the Barrens.

On foot, staggering under a sixty-pound pack, Hearne advanced into an area of some 500,000 square miles of open rock. He smeared his face with goose grease to keep off the maddening swarms of insects. They found no game to shoot, and survived on berries. Hearne refused to eat the lice or warble flies that

the natives plucked from their hair and clothes and cheerfully consumed. The party waded through swamps packed with hidden chips of rock that cut their clothes and skin to shreds, an experience they described as walking through porridge filled with razors. Hearne had heard tales of rich copper mines in the Barrens, and some of the Indians had copper implements. So on he went, attacked by storms and robbed by his guides, into an area larger than western Europe, where even the caribou suffer torments from the flies.

◆

The Europeans who followed Hearne did find copper in the north, and gold and silver and uranium. Like the explorer, they endured the cruel Barrens.

A tradition of the dauntless prospector runs through geology, and Blusson and Fipke fit the mold. When the joint venture that had funded them shut down, they collected themselves, raised money where they could, and went out to find diamonds.

Diamond indicator minerals lay widely dispersed in the Barrens. Glacial ice had collected the mineral grains from their source and carried them "down ice"—in the direction of glacial flow. Some of the grains were pressed into the ice itself, while others were flushed along by meltwater flowing under tremendous pressure beneath the ice. Compelled by these forces, the indicators often traveled tens of miles in a down-ice direction. Later, when the glacier melted away, trains of minerals remained behind on the surface of the land. In theory, the prospectors had only to pick their way back up ice to find the kimberlite source from which the grains had come. But glaciers are notoriously hard to track. They have slipped and slid in ways that are hard to reconstruct. Summer after summer the prospectors would isolate promising trains of indicators, track them back up ice, and find the trail go cold. Fipke sent garnets south to Houston, where Dummet ran them through Superior Oil's lab. That support ended in 1985, when Mobil Oil bought Superior and sold the diamond lab.

In the Barrens, year followed year, and still no kimberlite. Blusson began to concentrate his efforts elsewhere. Fipke formed a company, Dia Met Minerals, obtained a listing on the Vancouver Stock Exchange, and peddled the stock to friends at seventeen cents a share. His barber bought some, and the owner of the Greek restaurant in his hometown of Kelowna, British Columbia. Whatever he raised, Fipke spent in the Barrens. He became obsessively secretive, hiding the location of his prospecting even from some of Dia Met's directors.

After seven years of exploration, Fipke tightened his search to a zone of promising mineral trains on the Slave Craton, a 360,000-square-mile slab of Archean rock 2.6 billion years old. The target ground lay three hundred miles northeast of Yellowknife, the capital of the Northwest Territories, and just north of Lac de Gras, a fifty-mile-long stretch of stormy water. Low hills rose from the shore and boulders littered the tundra. Water and bare rock and the meadows of the tundra stretched away for miles in all directions. Fipke established a camp at Exeter Lake, north of Lac de Gras. As he sampled, he grew more excited. By now he had a lab of his own in Kelowna and, more important, had learned the secret of the G10 garnet.

Now, in the country north of Lac de Gras, Fipke found G10s in abundance. He staked a block of claims, concealing ownership by recording the claims in the name of a nominee. Nevertheless, Chris Jennings heard about it. He learned that Fipke's search had narrowed, and concluded that the prospector was close to a discovery. At this time Jennings was working for International Corona, a gold producer. He had left Falconbridge and a subsequent employer in frustration at being kept from the diamond hunt. Corona's directors themselves were never more than tepid about diamonds, but they had allowed Jennings to keep up a modest program.

In the summer of 1990, hearing that Fipke was staking more claims, Jennings took a gamble, putting $250,000 into a secret exploration budget. He then sent Leni Keough, a twenty-eight-year-old staff geologist, to sample down ice from Hudson Bay, sweeping across the whole of the central Northwest Territories from Hudson Bay to Great Slave Lake. Keough hired a floatplane and off she went. She stayed at fishing camps or native communities or camped on the tundra. She performed rudimentary geophysical tests and examined magnetic anomalies identified on government surveys. Mainly she sampled the eskers—gravel ridges formed from material washed out of melting glaciers.

"I had a set of field sieves with me," Keough said, "and sometimes at the end of the day I'd sieve the samples down to a concentrate and look at them through a microscope. Gradually I started to find indicators. I didn't even need a microscope. They were right there—beautiful purple garnets, chrome diopsides, ilmenites. When I started to see all these, I phoned Chris [Jennings] from a fishing lodge and told him that I could see the indicators, and he didn't believe me. He didn't think I could see them myself, just by looking at a raw sample."

In September, with the first snow swirling in the air, Keough arrived near Lac de Gras. She began to traverse the area in straight lines, collecting samples as she went. The days were shortening quickly, and the program would soon end. A rime of frost crusted the ground in the morning. The tundra scrub turned scarlet. One morning Keough began a traverse in country north of Lac de Gras, and came across a place disturbed by sampling. "I can't tell you now if it was Chuck's [Fipke's] dig, but it was definitely right around there, in that same area."

What Keough found on that traverse sent Jennings hurrying from Toronto to Yellowknife. The rich purple of pyrope garnets glowed in the gravel. Chrome diopsides were easily visible to the naked eye. Keough proudly displayed the grains. "Oh, we were ecstatic," she said. "I hadn't believed her," Jennings admitted, "and when I got there I found she was right. The samples were red with garnets." One sample contained microdiamonds. A microdiamond is a tiny gem, no longer than five millimeters in the longest dimension. Although they are small, they are very important. As geologists say, the best diamond indicator is a diamond.

Now the pace of exploration rapidly increased, because Fipke had found a microdiamond too. From his camp at Exeter Lake he sent out teams of dirt-baggers, as soil collectors are called. The dirt-baggers wore camouflage clothing to conceal themselves from planes. Their samples went to Fipke's lab in Kelowna. With a scanning electron microscope, Fipke could isolate and assess the G10s. But one important step was still beyond him: as powerful as it was, his microscope did not allow him to make crucial distinctions among the eclogitic garnets. Without this ability, Fipke could not predict the chances of a potential target's enrichment by high-grade eclogite. He turned to Gurney.

The kimberlite lab at the University of Cape Town had a microprobe, which could analyze the mineral chemistry much more accurately than Fipke's instrument. Fipke picked out the orange eclogitic garnets, fixed them to a cardboard sheet, and sent the sheets to Cape Town. Rory Moore, Gurney's partner, performed the final analysis of the eclogitic garnets. Once again, chrome was the element they measured. If the garnet was relatively high in chrome, then it came from inside the diamond stability field, and predicted an eclogite component in the source kimberlite.

The board of Dia Met began to press Fipke to stake more claims. The ownership of more ground would increase the value of the company if they made a strike. In Canada, mineral rights belong to the federal government. A licensed prospector gains the right to exploit minerals by staking a claim—hammering wooden posts into the ground at prescribed distances, and registering the claim with the local mining recorder. But claims are public information. Newly staked claims are posted in the recorder's offices. In Yellowknife, prospectors regularly check the recorder's notice board. If Fipke had staked a large block he might have found himself suddenly challenged by aggressive staking in the field, as speculators gambled he had made a discovery and hastened to acquire nearby ground. So he staked only a few claims at a time, and put it about in Yellowknife that he was looking for gold.

The ground was thick with indicators, yet there was no pipe. Then one day, as Fipke was flying over the property in a helicopter, he noticed a shape he had not noticed before. It was a small round lake. The shape reminded him forcefully of a pipe. Suddenly the realization struck him: it *was* a pipe. The pipe was under the lake!

With this sudden leap of perception, the problem of the hidden pipes resolved itself into a series of neat, logical steps. Kimberlite was soft. The surrounding granite, by contrast, was the hardest rock in the world. A glacier would have slid along and gouged the top out of a pipe and carried off a load of indicators, spreading them down ice. Thousands of years later, when the glacier melted away, water would have filled the scooped-out depression at the top of the pipe. The pipe was there, perfectly intact, under the lake. What is more, the country was speckled with small lakes. Fipke's own claims had dozens of them. He

Chris Jennings ice fishing in the Barrens. (Jennings family)

landed and hurried to the lake with his son. On the down-ice side of the lake, exactly where a glacier would have pushed it, Mark Fipke found a chrome diopside as big as the end of his thumb.

Fipke called Hugo Dummett, who had become North American exploration chief for BHP Minerals, a subsidiary of Broken Hill Proprietary Company, an Australian mining conglomerate. Dummett flew to Kelowna to study Fipke's data. Then the assessment of Fipke's indicators came in from Cape Town. The report stated simply that the data from Fipke's indicators were "the best for diamond potential that we have seen anywhere in the world, and we have no doubt that diamondiferous kimberlite is the source of the heavy minerals." John Gurney and Rory Moore were geology's leading diamond-indicator scientists, and they had virtually gushed. Dummett led BHP into a deal to spend as much as $500 million to develop a diamond mine. They drew a curtain of secrecy around the little lake and began to expand their camp.

By the end of 1990, masses of mineral evidence pointed to a diamond field on the Slave Craton. In one esker sample alone they found six thousand garnets. They hired Rory Moore and flew him in from Cape Town. Next they launched a program of airborne geophysics. Dummett wanted to find out how many other targets looked like the small round lake, because they would stake those too. To help on the ground, Fipke hired Ed Schiller, a short, irrepressible bundle of energy with the hammered face of a boxer. Schiller had lived in Yellowknife as resident geologist for the Geological Survey of Canada, and inevitably, with such well-known people now traveling back and forth to Lac de Gras, the prospecting community began to suspect that something important was happening.

"One afternoon I was down at the floatplane base," recalled Doug Bryan, a Yellowknife geophysicist. "We were watching Chuck Fipke and Eddie Schiller load sieves and some kind of milling equipment into an Otter. The word was out they were looking for diamonds, but frankly it just seemed ridiculous."

It did not seem ridiculous to Chris Jennings. He wrote an urgent memo to his board, pinpointing Fipke's camp and predicting that the area was on the verge of a staking rush, and that De Beers too would be involved.

While Jennings fretted in Toronto, Dummett told Ed Schiller to double the area of the claims. Schiller's staking attracted notice. "I'd been watching that block for years," said Bryan. "In typical gold or base metals staking, there's a rough logic to the way the block develops. There's a pattern where you follow stratigraphy. But this just seemed to grow in different directions at different times. I talked to my partner, but he just said, 'Oh, somebody's looking for gold,' which is what Chuck was telling anybody who asked."

By September of 1991, Stuart Blusson, rejoining the hunt, had found masses of indicators down ice from the little lake. Schiller remembers scooping up chrome diopsides by the handful. Dummett was now ready to put everything to the test, and drill the lake, the only way to know for sure what lay beneath the water. Fipke hesitated, and argued for a delay. He wanted to gather more mineral evidence before they committed themselves to drilling. He feared that if the lake was a dud, BHP would abandon the whole property, just as Superior Oil and Falconbridge had abandoned him eight years earlier. Dummett was adamant. He told the others to fly in a rig and drill the target before cold weather shut the camp.

Ice was forming at the edge of the lake when the drill arrived. A four-man crew accompanied it—two shifts of two drillers each. They would run the drill twenty-four hours a day. A helicopter slung the pieces of the rig into place at the northwest corner of the lake. No one told the drillers what the target mineral was. Even the name they gave the lake—Point Lake—was supposed to fool spies, since another Point Lake existed already and, unlike their lake, was named and identified on maps. If a driller on leave happened to gossip about Point Lake, competitors would waste their time flying off to the wrong lake.

Dummett and Schiller sited the drill. The engine roared into life, expelling a puff of black smoke from the exhaust pipe on the drill-shack roof. Fipke flew out of camp on business. Schiller settled into a routine, shuttling by helicopter between camp and drill. The drill passed from overburden into bedrock. It went day and night. Every day, Fipke called Schiller and asked the coded question, "How's the fishing?"

"Haven't caught anything yet," Schiller reported the first time Fipke called.

The next day Fipke called again. "How's the fishing?"

"Not a bite, Chuck."

On the morning of September 9 the drillers on the overnight shift noticed an abrupt change. They'd been drilling through hard black rock when suddenly the drill surged forward into something soft. They had no instructions to stop, and did not know what they were looking for anyway. They continued to drill. Schiller was at breakfast in the cook tent when the helicopter brought the night shift in. "Ed," the foreman said, "we've just got into the weirdest rock I've ever seen."

Schiller abandoned his breakfast, tore out of the tent, and yelled for the helicopter pilot. The flight to the drill took five minutes. When they landed, Schiller ran into the drill shack and opened the core box, where the drillers store the long cores of rock pulled from the hole. "The core was laid out in the box and I bent over it," Schiller said. "It had gone from a neat round core of dark rock to this greenish, gray, crumbly stuff. I got back to camp as fast as I could and put through a call to Chuck. When he came on the line I said, 'Chuck, we just caught the biggest fucking fish you ever saw.'"

They drilled the pipe to a depth of 950 feet. Fipke collected a 130-pound sample of kimberlite and flew to Kelowna. In his lab they crushed the kimberlite, then washed the soils away, then separated the light from the heavy minerals. When the process was complete, Fipke had a tray of heavy minerals. He snatched the tray and rushed it to a microscope. The diamonds were unmistakable among the darker stones.

BHP did not immediately release the news. The longer it could keep outsiders from learning of the strike, the more time it would have to extend its claims. Dummett increased the program of airborne geophysics. On the printouts, lake after lake stood out bright orange—a color enhancement that identified characteristics sharply different from the surrounding Precambrian rock. As they studied the results, the conclusion was inescapable: they had found an entire field of pipes. Two months later they issued a terse release, and the news went out that a tiny company called Dia Met and one of the largest miners in the world had just found eighty-one small diamonds in a lake.

5 Rush in the Barrens

In 1991, the year of the Point Lake diamond strike, De Beers's total sales of rough diamonds ran to $3.9 billion. Most of this came from its own mines in South Africa and Botswana, and from the lavish beaches of the Diamond Coast. Contractual purchases pulled in more tens of thousands of carats of top Angolan goods, millions of carats from the great pipes of Siberia, and a tide of cheap rough from the huge pit of Australia's Argyle mine. Other sources—dealers, traders, African rebels, thieves, and garimpeiros—sent a steady current of rough into the diamond quarters of Antwerp and Tel Aviv, where De Beers, although not the only buyer, was an active and important one.

The picture of diamonds, then, was of a monolithic industry, in which some 80 percent of the rough was gathered in by De Beers and channeled through its selling office in London. One might have thought that any new discovery would be destined, by a sort of law of gravity, to tumble into this gaping maw. But, in fact, there were reasons why the Canadian discovery might not, or perhaps *could not*, flow into the De Beers stream, and those reasons were bad news for the cartel.

For one thing, BHP was a large, international mining company. Its executives had plenty of experience mining and selling minerals. They would not easily accept the assertion that another company knew what was best for the commodity they had found, including the price that should be paid for it. If the Point Lake discovery should prove to be important, De Beers might soon face a rival as burly as itself. As mining observers liked to say, one eight-hundred-pound gorilla had met another.

A second factor that could prevent the sale of future rough to De Beers was BHP's involvement in the United States. The BHP subsidiary that held the diamond interest was headquartered in San Francisco. Among its assets were coalfields in the American Southwest. There was speculation that those interests could be jeopardized by any deal with De Beers, because of a history of antitrust action against De Beers by the United States Department of Justice, in which the company had faced charges of conspiring to fix prices.

Indeed, De Beers has not answered the latest charges, brought in 1994, and its failure to do so has produced a bizarre result. Because the charges are outstanding, De Beers executives do not travel to the United States, where they might face subpoena. The most senior managers of the world's preeminent diamond company are thus effectively prevented from setting foot in their largest market.

Because of this strange legal quagmire, the Australian company, with its own interests to protect, might therefore have been chary of a deal with De Beers.

But the aspect of the discovery most antagonistic to the cartel was the fact that it had happened in Canada, where the mineral exploration scene is dominated by a host of small, unruly companies, called "juniors." Juniors raise exploration money on

Robert Gannicott. (Aber Diamond Corporation)

the stock exchanges. Investors call the issues "penny stock," because they sell for pennies a share. The buyer of such stock hopes that a mineral strike will convert the pennies into dollars, for windfall profits. Always on the lookout for the latest opportunity, juniors descend upon new discoveries in a mass. The idea of yoking such a promoter-driven and combative group to some larger purpose, such as commodity price control, would be laughable.

Turbulence aside, the junior market has at its heart the powerful ideal of discovery and a corps of seasoned explorers. A group of these happened to be meeting in Toronto when news of the diamond strike broke. Among them was a pair of close friends and business associates, Robert Gannicott and Grenville Thomas, both veterans of the exploration game and well suited to the events that followed.

Thomas is a debonair and popular figure, always and everywhere known as Gren. He has a quiet voice and a smile often forming at the corners of his eyes. He grew up in the Welsh coalfields, and at the age of sixteen started work as a miner himself.

But he studied at night, earned a degree in mining engineering, and in 1964 moved to Canada. Gannicott is a bluff Englishman, with a history broadly similar to Thomas's. A native of Somerset, he came to Canada in 1967 as a teenager and worked in the northern mines. He met Thomas there, and the two became close friends. After a few years Gannicott left the north to study geology at the University of Ottawa, graduating in 1975. Both men had worked for large mining companies, and both now ran juniors.

The meeting in Toronto ran through its business, then turned to a discussion of the diamond discovery. "We didn't have the faintest idea what to make of it," Thomas recalled. "No one knew anything about diamonds. We were sort of sitting there going, So, okay, these guys have found some microdiamonds. What the hell does that mean? Who do we know who knows anything about diamonds? And somebody said, Chris Jennings. Jennings knows all about diamonds."

They tried to phone Jennings from the boardroom, but could not find him. In fact, he had quit his job again, and was in London trying to raise money for a diamond exploration company of his own. When they finally tracked him down and asked him about the news, Jennings was astonished. He had not heard it. He took the first plane back to Toronto and joined the others as soon as he arrived.

"Well, he had it all," remembered Gannicott. "He came in, and we had a meeting, and he poured it out. He had the whole thing down pat, why there were diamonds there, why it was going to be the greatest diamond strike since Kimberley. You know Chris—*bigger* than Kimberley. By the time he was through we could hardly wait to get up there."

The partners quickly worked out a deal in which Jennings would provide the group with advice on where to stake in return for a royalty on any future production. He would also receive shares in the company that would hold the staked ground. For that purpose, the group settled on a private company they owned, which had been established to explore for gold in Iceland

Gren Thomas (Gren Thomas)

and still had a few hundred thousand dollars in the bank. With the deal done, Jennings and Thomas immediately left to pack. They needed to move fast. Winter was closing on the Barrens. Daylight in November lasted six hours, and would decrease by five minutes a day. Jennings and Thomas flew to Yellowknife on November 20. The capital glittered in the clear arctic night—a small city of ten-story office buildings, blazing with light, and prosperous suburbs etched into the thin forest.

Thomas and Jennings separated as they came off the plane. Each went through the airport by himself. They were well known in the mining community and did not want people to see them together. The diamond strike had not yet ignited the town, and Jennings and Thomas were determined not to give it any reason to do so. "We thought the whole place was just trembling on the edge of a staking rush," Thomas says. "Here was this news about diamonds, and we had heard from some pilots we knew that De Beers was chartering helicopters, and we expected the whole place to blow wide open."

They made their way out of the airport into the frigid night. The thermometer stood at twenty-five below. Snow blew back and forth in the streets. In separate vehicles they drove to a cheap motel, where it was unlikely they would meet anyone they knew.

Waiting for them was Doug Bryan. He reported the steps he had taken toward establishing a winter tent camp in the Barrens. They would need a helicopter for winter staking, and one would fly out the next day. A staking crew stood by in Yellowknife. Thomas and Jennings went over the plans: Thomas would remain in Yellowknife for a day to organize support, then join the camp. They reviewed the maps of the area they hoped to stake, then turned in for a fitful sleep.

Two hours before dawn Bryan arrived back at the motel. Jennings transferred his gear to his truck and the two men drove through the dark city and made their way to a private hangar at the airport. A twin-engine plane was running up its engines; the helicopter stood nearby. Two stakers were already waiting, and everyone helped load the plane, piling in lumber and tents and food. Last they rolled drums of aviation gas up the ramp into the cabin. The helicopter would consume a lot of fuel, and all of it had to go in by plane.

In half an hour they were ready to go. Bryan gave the helicopter pilot the coordinates for Gloworm Lake. The helicopter rose from its pad and clattered away to the northeast. The others clambered into the Twin Otter aircraft and the plane taxied out. A landing rig fitted with a combination of skis and wheels would allow the plane to take off from Yellowknife's cleared runway and land on snow when it reached the frozen lake, two hundred miles away.

A murderous wind was sweeping down the length of Gloworm Lake when they arrived. The pilot circled the bay where the camp would go, and studied the pattern of snowdrifts on the surface. He came down at the south end of the bay, tossing up bursts of powdery snow as he taxied in. The helicopter had already arrived, and the pilot was tying it down in the shelter of an esker. The plane roared up off the lake and slewed to a halt near the helicopter. A watery dawn had just begun to illuminate the land. The temperature was thirty below.

Their first priority was shelter. The cells of human skin are filled with water. In very cold weather this water freezes and ice crystals form inside the cells, damaging them. This condition is called frostbite. In the wind at thirty below, exposed skin will freeze solid in half a minute.

They wrestled a wooden floor into place and put up the tent frame. Gusts hacked at the canvas as they grappled the edges and dragged it over the frame. With sledgehammers they drove the tent pegs into the frozen ground. They hauled in an oil stove, fed the stovepipe through a collar in the roof, and got a fire going. In an hour they were done—insulation tacked into place, food stacked inside and out, staking posts piled in the snow. A generator ran to keep the helicopter warm. The Otter taxied down onto the lake, put its nose into the wind, and roared away.

Inside the tent the oil stove muttered and hissed. A small balloon of oven-hot air expanded around the stove; the rest of the tent was freezing. The stakers dragged in posts and began to hammer on the regulation metal tags. The claims would be staked according to strict rules laid down to prevent disputes: Tagged posts would be hammered into the ground at specified points, and the resulting claim would be duly recorded with the government mining recorder in Yellowknife. As the stakers worked on their posts, Jennings spread out a staking map and reviewed the ground. Fipke and Dummett had staked more claims around their discovery, creating a large block of protected ground: the discovery block. This big block was hatched boldly onto the staking map. Jennings tapped his finger on the paper at a place just north of it, indicating the area where Leni Keough had found the richest trains of indicators. That would be their first staking target.

In the morning the stakers loaded their posts into the helicopter and headed north. Winter erases the surface features of the Barrens; visible landmarks disappear. The helicopter flew by global positioning system (GPS), which relies on satellites to furnish navigational coordinates. Soon they approached the edge of the target block. As soon as they flew in they saw fresh posts hammered in the snow. Someone had gotten there first and taken the ground. Bryan thought the posts might be Fipke's. They landed and climbed out and inspected the new posts. They were not Fipke's. The posts were De Beers's.

"We flew around until we found the southern boundary of their staking," Bryan said. "There was one little gap they'd left unstaked, about thirty thousand acres, and we grabbed it. But all that other ground north of BHP, De Beers had nailed it all down tight."

When they got back to Gloworm Lake with the news, Jennings accepted it calmly. He said he was not surprised, he'd expected them to show up. He unfolded the big staking map and brushed his hand over the territory east of the discovery. "We'll take the up-ice ground, " he said, "all of it."

At nine o'clock the following morning, a weak sun rose a few degrees above the horizon and cast its thin light on the Barrens. A storm was moving down from the arctic coast, and the first gusts had begun to racket against the tent. The helicopter whined into life and the stakers piled in with posts. The machine rose up in a cloud of snow. They flew away to the west until they reached the eastern shore of Lac du Sauvage, which formed the eastern boundary of BHP's staked ground. They began to peg out the first claims. They took sixty thousand acres in a day. That evening Gren Thomas arrived at Gloworm Lake to help with staking. Then the weather struck.

The storm came down with howling winds and sealed the camp shut. The men battened down the tent and waited for the wind to fall; instead it rose. The daylight hours passed. The stakers played cards and Jennings and Thomas went over the maps. At six o'clock they undid the flap and someone crawled outside with a pick and chipped a few steaks from a frozen stack. After supper they read or tried to sleep, while the generator hummed against the storm.

Next day the wind rose again. The sides of the tent thundered and volleyed. Jennings lay in his sleeping bag, trying to read. At nine o'clock a feeble dawn arrived, and the pilot crawled outside to inspect the helicopter. The shore of the lake, twenty yards away, was invisible from the tents. The helicopter rocked against its ropes, and the tent walls snapped and cracked in the demented gusts.

For an hour the gale rose. It howled across the esker and through the anchor ropes. Conversation became difficult. The stovepipe rattled in its collar. Then, suddenly, a powerful back-draft blew down the chimney and snuffed out the fire. Soot erupted from the stove, blinding the men and clogging their throats. They gagged and retched, and the temperature plunged. Jennings groped for a sweater but couldn't find it. Finally they managed to get the stove going again. The temperature crept back up and Jennings and Thomas set to work cleaning the soot from the maps.

Next day the winds fell. Blue sky appeared. With the sun came a drop in temperature to forty below. The helicopter lifted off and headed for the staking ground. When they reached the corner of the first claim, the helicopter touched down. A staker leaped out, drove in a post, and threw himself back into the helicopter, which immediately took off and charged along to the next point where a post must go. Stakes went into the snow every fifteen hundred feet, enclosing claims of twenty-five hundred acres each.

By noon the weather had begun to deteriorate. A second helicopter arrived to help with the staking. The pilots were worried about whiteout—a condition in which blowing snow obliterates the ground and pilots can quickly become disoriented. They decided the conditions were just stable enough to keep flying. Then the GPS began to fail. Bright green coordinates would be scrolling onto the display when suddenly the screen would go dark. When it did, the staker would unfold his map and stare out at the shifting white world and try to identify some feature, indistinct and partly glimpsed, that matched a topographical detail on the map.

The crews began to take shortcuts with the staking rules. Posts are supposed to be driven into the ground, but in what came to be a common practice the pilots fitted their craft with small, hinged windows, and when the crew thought it had reached the right place, the staker would flip the window open and shove out a post. They called it "airmail staking." If someone later contested the claim on the grounds of improper staking, the claim owners could always suggest that a bear must have knocked out the post.

Between storms, the staking maps were filling up with claims. Soon they had 500,000 acres. On one day alone, one of the helicopter teams staked a block that ran for eighteen miles along one side. They took almost ninety thousand acres close to Gloworm Lake itself. Someone discovered an unstaked block of fifty thousand acres south of the BHP discovery claims. The whole block lay under the waters of Lac de Gras, but they took it anyway. Then the money ran out.

The company's treasury had evaporated into the bills for helicopters, fuel, and men. The partners knew there was only one place to get the money to continue staking: the stock market. But the venture rested in a private company, which had no listing on a stock exchange. They needed a public company, and luckily one of them had one: Gren Thomas's Aber Resources traded on the Toronto Stock Exchange. So the partners transferred the claims owned by the private company into Aber Resources. As payment they took Aber shares. Aber now owned the ground. Notices went out to the financial markets that a public company with a huge land position near a diamond discovery had shares for anyone to buy. Yet no one did.

"It was scary," Thomas said. "We were all waiting for something to happen, and nothing happened. We expected Yellowknife to go wild. There should have been staking all over the place. We thought there would be helicopters heading up to the Barrens in droves. There wasn't. Yellowknife was totally quiet. It was as if nobody'd even heard about diamonds. It was like a phony war. De Beers had staked their ground, and we had ours, and nobody gave a damn. Things just died. They thought it was all a scam. I thought, God, we've staked all this ground and nobody thinks there's anything there. You start wondering if maybe they're right."

At this low moment Thomas's daughter, Eira Thomas, a newly graduated geologist, came to work at Aber. There were few field positions for young geologists at the time, and Eira Thomas

Eira Thomas. (Matthew Hart)

gladly accepted an office job at her father's company. Every morning she reported to the ninth floor of the old Marine Building on Vancouver's waterfront. Documents spilled from cardboard boxes. Aber's claims in the Barrens had grown to a million acres, and page by page, Eira Thomas put it all in place. Tall and thin, with blue eyes and a calm demeanor, Thomas subdued the chaotic evidence of other people's fieldwork. She looked like an explorer herself, in jeans and a loose sweater, with her waterproof jacket tossed in a corner. Sorting through the papers, she could not help but recall her summer jobs in the Barrens, hiking out onto the tundra with people like herself. Still, she did not think she would escape the drudgery very soon. As a geologist she knew that the chances of finding a pipe were slim. The market seemed to agree with this assessment.

It was January 1992, only two months since the partners had flown to Gloworm Lake. They had made an "area play"—staked ground in the area of a discovery in the hopes that the market would judge the claims to be of value. But the market showed no interest, and Gren Thomas and Robert Gannicott decided to look for a partner, a large mining company that would make a cash investment in return for a share of the claims. They settled on the British company Rio Tinto, the largest mining company in the world, and negotiated a deal. It turned out to be a very good deal for Rio Tinto, because the ink was barely dry on the agreements when the diamond play erupted. It was as if some force had been gathering out of sight, building to a critical mass, and now it arrived with a rumble and a roar. Investors awakened to the discovery. The shares of Charles Fipke's Dia Met climbed to C$1 a share, then C$6, then kept going.

◆

In late January 1992, two months after BHP and Dia Met announced their discovery, Dia Met's share price passed eight dollars. Aber Resources, the newcomer, followed suit, jumping from twenty-five cents a share to C$1.35. *The Northern Miner* reported that 1.6 million acres had been staked around Exeter Lake. But as soon as this appeared it was out of date. De Beers alone had staked a million acres. The discovery block itself—the core block of claims on which the original discovery had been made—now comprised 850,000 acres. Fipke christened these claims the Corridor of Hope, and put it about that claims staked outside the Corridor of Hope would be barren, a prediction that quickly became known as Fipke's Curse. But nothing could stop the rush.

In Yellowknife, Great Slave Helicopters expanded its fleet from thirteen aircraft to thirty as junior exploration companies hastened north to join the rush. Pilots were hired from as far away as Australia. Sheets of tin prospecting tags went out of the mining recorder's office in stacks—eleven thousand tags in February alone. A cavalry of helicopters ranged across the Barrens with staking crews. Juniors in the play staked claims from Great Slave Lake to the Arctic Ocean. Traffic increased on the winter road, the seasonal ice highway that services distant mine sites in the Barrens. Pilots could see convoys thirty miles away—lines of headlights snaking through the winter night. BHP was hurrying supplies up to its expanding camp, and every new truckload increased the speculation that now drove the play.

A diamond-exploration camp at the edge of the Barrens.
(Matthew Hart)

Ed Schiller broke away from Chuck Fipke's company and optioned 125,000 acres of his own at Yamba Lake, northwest of the strike. Chris Jennings too struck out on his own. He assembled 700,000 acres of claims, and bought a dormant listing on the Toronto Stock Exchange called SouthernEra Resources. He sold stock for one cent a share. In a few months SouthernEra was trading at C$1.90—a return on invested capital of 18,900 percent. Aber's share price rose to C$2.34.

All this moved on the Barrens like a wind, blowing a cloud of helicopters and men. Staking posts rained into the snow. The equinox came and went, and the hemisphere tilted to the sun. The arctic summer came on in a rush, adding an hour of daylight every ten days. Tents sprang up beside the gravel beaches. Eira Thomas stuffed her last sheaf of papers into a filing cabinet and headed for the Barrens too. Flush with the fortunes of a rising market, Aber Resources was preparing to explore its claims.

In May, Aber moved the camp from Gloworm Lake to Lac du Sauvage. The site faced a broad bay with a wide sand beach where floatplanes could run up on shore. In late May the water was still clotted with melting ice, so they cleared a landing strip on an esker behind the camp. The exploration team arrived in Yellowknife; Lee Barker, an Aber director, was in charge. The others were Eira Thomas, Leni Keough, an American diamond geologist named Mike Waldman, and Bjarke Schönwandt, a Danish geology student whose father was a friend of Robert Gannicott. Eira Thomas's luggage included Thor, a sixty-pound northern sled dog, part wolf, with white hair and one blue eye and one brown. Thor liked planes, and as the geologists loaded their gear into the chartered Otter, he scrambled in and took his place among the packs. An hour and a half later the plane dropped down onto the esker and bumped to a stop. A helicopter waited to ferry their equipment down to the camp. Thor leaped out of the helicopter, charged around, and pelted off into the tundra.

The camp had nine tents—office, cook tent, warehouse, lab, combined laundry and shower tent, and four sleeping tents. The tents were the standard white canvas "prospectors' tent" of the North. Each was the size of a cabin, erected on a wooden frame, insulated, floored, and fitted with a proper door and an oil stove. There were raised platforms for the sleeping bags. From the tents, a flat shelf of tundra ran down to the beach, a quarter mile of unblemished sand.

A long, thin esker ran in a straight line from one end of the beach into the lake. The cold water teemed with trout. A few of the explorers cast from the beach, and the cook built a fire with green arctic willows in a little smokehouse beside the kitchen and put the fish they caught on racks inside. That first evening the geologists gathered in the cook tent and ate smoked trout.

On Eira Thomas's first day a helicopter dropped her at the northern end of Lac du Sauvage. Twenty miles away, an Otter glinted in the sun as it made its approach to the BHP camp. Thomas turned and walked east over a hill until she found the rocky course of a stream. She slipped out of her pack and began to dig. Summer was short in the Barrens; everyone understood the need to sample widely and quickly. The days fell into a routine of long, exhausting work.

The Barrens crackled with excitement. BHP found two more diamond-bearing pipes. Stock market analysts and promoters, inevitably, began comparing the Barrens to South Africa. Every garnet seemed to call for a press release. The geologists worked through the long daylight hours, then sprawled by the shore and watched the evening fade. At midnight they made their way back to the tents and fell instantly asleep.

One night Thomas and Keough woke to the sound of Thor whining loudly in the tent. Keough stuck out her head and saw a grizzly wandering through camp. "Thor was frantic," Thomas said. "He tried to burrow into my sleeping bag. I ordered him to go outside, but he wouldn't go. Finally Leni and I managed to push him out and fasten the door behind him." Instead of chasing the bear, Thor wailed in terror and dashed to Schöndwandt's small, single-person tent. "He almost tore it apart getting in," said Thomas. The bear panicked at the commotion and galloped out of camp. Not every encounter with a grizzly ends so well. The bears' keen smell draws them to prey across long distances. They can appear suddenly loping over a hill, an animal the size of a small car moving at the speed of a cantering horse. In the Barrens a lake is always near. Panic may send an undefended person rushing into the water to escape attack, and the lore of the North abounds with stories of people who have died of hypothermia while a grizzly bear paced to and fro on shore.

◆

As the Barrens swarmed with explorers, pressure grew on the geologists to produce news that would distinguish their claims from those of competitors. But diamond geology was new to most of them. BHP had diamond stars like John Gurney and Rory Moore on the payroll, and with known pipes in hand, knew what to look for. Other than De Beers, no explorer in the Barrens could match this expertise. It was in an effort to close this knowledge gap that Aber risked the ire of BHP.

Aber's geologists did not know for certain what a diamond pipe in the Barrens should look like on the printouts from the airborne surveys. The color-enhanced printouts looked like maps of molten galaxies: huge splotches of red and green and blue squirming across the plots. But which was a pipe? They needed a known pipe to provide them with the key. There was only one location in the Barrens with known pipes—the discovery claims. They decided to fly over the Point Lake pipe and collect its signature.

With the pod of instruments fastened to a tether, the helicopter lifted away from Lac du Sauvage. It dipped across the eastern end of Lac de Gras, cut sharply inland, and clattered across Point Lake. Men rushed out of the trailers and fixed the intruder with binoculars. The pilot completed his run, turned back, and crossed the pipe again. The agitation of the men below was clear. In a few minutes it was over, and the pod of instruments, with the geophysics of the little pipe recorded, went trailing back to camp behind the helicopter. An angry letter from BHP arrived on Grenville Thomas's desk in Vancouver, but after all, the geophysics of a country is no one's property, and the incident faded into that general stock of acerbity that vitalizes every mineral play.

Aber's mission that first summer was to locate likely targets to be drilled. While wide sampling for indicators went on throughout the claims, the crucial airborne work unraveled the changes wrought in the craton over tens of millions of years. Airborne geophysical surveys measure many properties of the rock, including magnetism. All rocks are magnetic to some degree, and this magnetism can be measured from the air with a magnetometer. The granite of the Slave Craton returns a flat magnetic signature, about one thousandth as strong as a refrigerator magnet. Younger intrusions into the craton, such as kimberlites, produce slightly higher readings. Sometimes, though, these differences in magnetism are so small they are undetectable, and so, in addition to simple magnetism, airborne geophysics measures electromagnetism.

An electromagnetic survey (EM) gauges the resistivity of the rock—the degree to which it resists a current of electricity. A transmitter in the instrument pod sends electric currents into the ground at different frequencies. When the signals come back, the EM unit records the decay, or weakened strength, of the returning signal. Anomalous data stand out, betraying new rock that has intruded into the host.

Another clue that newer rock has penetrated older rock is the direction of magnetism. Unlike true north, which is located by definition at the north pole, magnetic north wanders over time. Magnetic north and true north are never the same. Magnetic north changes its position in response to constantly changing flows of electricity in the Earth's core. These changes affect the strength and direction of the magnetic field at the surface. When a new intrusion of molten rock penetrates the Earth's crust, the new rock will be magnetized according to the position of the magnetic pole at that time. Some of its mineral grains, those with a strong magnetic susceptibility, will align themselves toward the pole. When this rock cools and hardens, those grains will be fixed as they were—literally frozen in time. The direction in which they point will forever distinguish them from the surrounding rock, and geologists can detect this with a microscope.

Every day the geologists learned more about the claims. On the ground they sampled widely, and one day made a startling discovery—someone else's posts. Checking the tags, they saw that the posts were Fipke's. He had staked one whole block of their ground, and staked it before them. And yet, for some unknown reason, he had not registered the claims with the mining recorder. "We didn't even know he'd staked there until the snow melted and we got that summer program going and went in and found his posts," said Gren Thomas. "I don't know why he didn't record them. Maybe he ran out of money, or he thought he had enough ground. He didn't contest our staking, there was no fight about it. But I don't suppose he liked it."

In the scramble of the staking rush, remarkably few disputes arose. Competing geologists sometimes met in Yellowknife to notify each other of staking targets, so as not to stake over each other's ground. Yet a tradition of blithe deceit has characterized Canadian prospecting and supplied many of its heroes with delightfully murky reputations. The stories are told of specious staking maps deliberately left in bars. Prospectors will allow foes to "eavesdrop" on the most intoxicating rubbish. They steal core from competitors' drill shacks and pilfer soil at night. They scrutinize each other from the air. This is the nature of a mineral play, and it supplied the context for the events that began at Lac de Gras on a certain summer's day in 1992 when a pair of Aber geologists went up on an esker, set up targets, and started to shoot.

"Thor took off down the lake," Eira Thomas remembered. "Guns terrified him. He was gone for three days. I thought for sure he was toast, that wolves had got him. Maybe he'd followed a female and she'd led him back to the pack and that was the end of him. He'd run with wolves before, but you never know with wolves."

On the fourth day a radio call came through from Fipke's tent camp at Exeter Lake. They asked if anyone was missing a dog. When the reply was affirmative, they asked if the dog looked like a wolf. They were told that, yes, he did.

Spooked by the gunfire, Thor had gone pelting up the eastern side of Lac du Sauvage and kept on running. He ran on and on until he reached the northern end of the lake, rounded it, and started down the other side. By the time he reached Fipke's camp he had traveled sixty miles. When they saw him coming along the shore, long-legged and loping, they thought at first he was a wolf. But wolves are afraid of men, and by the time Thor got to the tents they realized he was a dog. Exhausted and hungry, he headed straight for the cook tent. The camp made a fuss over the personable dog with the blue eye and the brown, and Thor settled in. Naturally, they wondered where he'd come from.

Thor wore a tag issued by the city of West Vancouver. Fipke's camp called the municipal licensing branch and asked if a dog with Thor's license number was reported missing. The municipality checked its records and called Eira Thomas's mother and asked if the dog was lost. She told them Thor was up in the Northwest Territories with her daughter, Eira. The city passed this information back to Fipke's camp, which received it with disappointment. Thor had eaten his way through an impressive share of their provisions and into their hearts. Reluctantly they called the Aber camp.

Thomas offered to jump in a helicopter right away and come to fetch the dog. The Fipke camp refused, apparently suspecting that Aber had set the whole scheme up as a means to get a closer look at its competitor's camp. Instead, they offered to keep the dog. When Thomas insisted she wanted him back, they flew him out to Yellowknife. The Fipke camp was a five-minute helicopter ride from Aber, but Thor made a four-hundred-mile round-trip to come home. The incident was reported gleefully on national radio, where Thor enjoyed a moment of fame as the Spy Dog of the North.

The summer of 1992 ended ecstatically for the diamond players. BHP sent out a release announcing the discovery of nine new pipes. The news landed in an eager market. A group connected to a European bank speculated in a report to clients that the Canadian diamond field might rival Botswana. Dia Met's share price climbed to C$20. One month later Aber drilled eight targets and found seven pipes. Delirium swept the market. Money poured into the play. A report from a London brokerage put the value of the discovery claims alone at C$2 billion. Inflated by this helium, market analysts asserted that the diamonds in the ground at Lac de Gras were worth C$9 billion, and that Canada would go on to produce one third of the world's annual gemstone diamond production. This claim marked the high water of the hype.

In 1993, the following year, the need to explore the targets took over, and some of the jubilation drained from the play. The euphoria of the staking rush was over. Now, as month followed month, investors reminded themselves that the odds of a pipe's containing an economic diamond lode were long. Moreover, BHP was pulling away from its rivals in a binge of discovery and construction. Its camp was large and bustling, and it had begun the laborious, crucial process of extracting large amounts of kimberlite and counting the diamonds to assess the viability of a mine. Compared to this grand project, the exertions of the juniors paled, and Gren Thomas considered how to strengthen his company's position.

He decided to attempt a merger with a group of juniors known by the acronym DHK. While Aber had a profusion of pipes, DHK had a single target, with the important distinction that the DHK property had already yielded the most promising indicator of all—diamonds. Rio Tinto was DHK's senior partner, and was now besotted with the juniors' target. Aber, it seemed, had fallen somewhat in the shadows. Compounding its problems, Aber's senior geologist quit and went to work for Chris Jennings, who was allied with Rio Tinto and the DHK group. Into this trough in Aber's fortunes stepped Eira Thomas, at the age of twenty-four, appointed chief geologist.

◆

In March 1994, after a brief sojourn in Vancouver, Eira Thomas was returning to Lac de Gras. Aber Resources had moved to a camp on the winter road. As Thomas looked out the aircraft window, she saw the headlights of a convoy winding from lake to lake. Traffic on the road was heavy that year, with BHP moving construction material up from Yellowknife. Just to the north, the lights of the sprawling BHP camp sparkled in the black expanse of night. Then the plane banked away, and the pilot lined up on a pattern of flares and came down on the lake. A cloud of snow churned up in the backwash of the propellers as the captain reversed pitch to brake the plane. They taxied off the ice and up a slope, and came to a halt beside the tents and steel trailers of the camp.

On that black winter morning a single target occupied

Thomas's mind: a geophysical anomaly called A-21. It lay under Lac de Gras, fifteen miles west of the winter road. A stream of indicators on a nearby island pointed to an origin in the lake, where the target lay. But before she could drill it, Thomas had to face a threat that was hanging over her newfound independence and, as she saw it, over Aber too—the merger with DHK.

Rio Tinto's passion for the group's target had only grown. The property was a pair of linked pipes beneath a lake called, in the local Dogrib language, Tli Kwi Cho, or "dog's balls." There were microdiamonds in the pipes. So impressed was Rio Tinto with the initial sampling that it had elected to proceed directly to a large bulk sample of five thousand tons of kimberlite, instead of taking the normal, more prudent course of advancing from small sample to larger sample in measured stages. The popular opinion was that Rio Tinto was anxious to catch up to BHP, which had already built a large processing mill to crush kimberlite and extract diamonds. In Rio Tinto's headlong rush to close its rival's lead, the company had flown the pieces of a processing mill to Yellowknife, reassembled it there, and rushed ore trucks up the winter road to gather kimberlite from Tli Kwi Cho. The stock market bet that Rio Tinto knew what it was doing, and investors bid up the stock of the juniors forming DHK. By contrast, Aber's share price lagged. That is how things stood that spring when Gren Thomas made his bid to merge with DHK.

Arriving at the camp on the winter road to discuss the proposal were two of the largest shareholders in DHK. Eira Thomas did her best to impress them with the potential of A-21. She had a sample of G10s collected from the island. But the rival explorers scarcely listened to her, and dismissed her samples. They pointed out that her target was under the waters of an enormous lake, one that could not be drained for mining as smaller lakes could. They left without accepting the proposal for a merger.

Now the shape of a race became clear to Eira Thomas, a race she would have to run to preserve her father's company. The drift of sentiment at Lac de Gras, among those associated with Rio Tinto, favored a merger of Aber with the others, and Thomas

The winter road. (Matthew Hart)

knew her father would continue to pursue it. She also knew the merger could go ahead only on terms disadvantageous to Aber, because DHK had all the momentum of discovery and market speculation and the hopes of Rio Tinto buoying them up, whereas Aber had only its targets, deep under water, and the conviction of a twenty-four-year-old woman with a handful of garnets.

Eira Thomas's brother-in-arms at Lac de Gras was Robin Hopkins, like Thomas a young geologist thrust into the middle of the hottest mineral play in the world. They laid their plans. They would drill A-21 twenty-four hours a day, in two 12-hour shifts. A helicopter would shuttle the crew change in and out. Thomas flew out to the island and sited the drill hole. A helicopter slung the pieces of the drill into place on the ice. They built the drill shack, and in the early hours of April 27, sent the drill down into the frigid depths of Lac de Gras.

At first the thick layer of packed mud on the lake bed fouled the drill. It took hours to force the bit through the mud. Beneath the mud lay gravel interspersed with boulders. Thomas and Hopkins returned to camp. They went back to the drill twelve hours later when the helicopter carried out the change of crew. In the meantime the drillers had pierced the gravel overburden and pen-

Using a magnetometer to pinpoint a target pipe at Lac de Gras. (Aber Diamond Corporation)

etrated the target. Now they reversed the drill and began to pull up the core, through the gravel overburden, through the mud, through the waters of Lac de Gras. When the muddy, dripping core came up into the shack, Thomas and Hopkins grabbed it from the drillers and rushed outside. Their spirits leaped as they examined it—a dark and crumbly, beautiful, messy, sodden length of kimberlite.

But was it diamond-bearing? The drill had barely tapped the topmost yards of kimberlite. The two geologists would have to wait for a better sample. They returned to camp with the outgoing crew. They passed the hours in the computer tent, examining the remaining targets. When they could wait no longer they jumped onto snowmobiles and tore down the hill onto Lac de Gras, roaring away across the ice.

It was a twenty-minute ride to the drill. When they got there, they ordered a "pull" of core. The drill had entered deeper, firmer kimberlite. Thomas and Hopkins took a length of core outside and crushed it with a rock hammer. They slopped some water into a gold miner's pan and sluiced the pulverized kimberlite around, floating off the lighter minerals and tipping them out of the pan until they had a residue of heavies in the bottom. They spread the minerals out with their fingers and saw the deep purple of pyrope garnets. The sample was loaded with indicators.

Over the next three shifts of drillers, some two hundred yards of core accumulated. As each pull came up, it was carefully placed in the long, open-topped, wooden core boxes. The geologists strapped it snugly into place, lugged the core outside, and lashed the boxes to a wooden sledge hooked to a snowmobile. Load by load they towed the entire sample back to camp. When they had it all safely in place, Thomas and Hopkins closed the door to the core tent and began to log the core, describing the appearance of each length. They split the core in half lengthwise, a standard way to select a representative sample for shipment by air while lightening the load by half. Each time they split the core they found more indicators. They worked through the whole night, their excitement mounting as they finished one length of core and started on the next. They tried to find a microdiamond but gave up. "We decided we probably wouldn't recognize one even if we saw it," Hopkins said. "Most of the things we picked out as possibilities turned out to be transparent Muscovite micas."

In the end they selected twenty kilograms of split core, put it in pails, and shipped it off on the next bush plane to Yellowknife. From there it would be rushed to a lab in Ontario, to be boiled in acid in huge kettles—a process called caustic fusion, which destroys the softer rocks and leaves the heavy minerals. When the plane left with the kimberlite, Thomas called her father.

"Dad, where are we with the merger?"

"Why?" asked Gren Thomas.

"Because I like this hole, Dad. We've got a great hole here."

As they waited for the lab results from Ontario, Eira Thomas and Hopkins noted with alarm that the daytime temperatures were swiftly rising. Sometimes the thermometer registered as high as zero on the Centigrade scale. The ice began to loosen along the shore, breaking into the thin, vertical shards that northerners call candles. The familiar tinkling of candled ice drifted through the Barrens as lake after lake began to stir in the sun. The chiming of the candles heralded the approaching end of the ice as a useful platform. Thomas decided to drill another target while she could. The drillers took apart the rig and moved it off to the north.

Two days later the anxiously awaited call came in from Ontario. Technicians had retrieved twenty microdiamonds from the sample. In the dry prose of her weekly notes Thomas wrote: "It appears that there is a strong likelihood that A-21 is indeed the source of the kimberlitic indicator minerals recovered from the [summer] till samples. The chemistry suggests that it should be diamond-bearing." The chemistry suggested it, the market inhaled the news, and Aber's share price jumped from C$4 to C$6 in a single day.

Now an acute dilemma faced Aber's chief geologist. The merger with DHK still loomed as a possibility. The A-21 sample results would not keep the market's interest in Aber afloat for long. Only a larger sample of kimberlite would establish beyond doubt the value of the pipe. But the days were growing dangerously warm. Snow had started melting on the tops of hills and on the south-facing slopes; the surface of Lac de Gras was softening into a spongy texture. Thomas and Hopkins surveyed the situation and made two crucial decisions: they would rush a heavier, wide-diameter drill onto A-21 to extract a larger sample, and at the same time they would pull the lighter drill off the midlake target they had moved it to and bring it closer to shore, onto what looked like a twin-pipe target.

The drillers balked. Water was already forming on the ice. They did not want to drill at all, let alone with a heavier rig. Thomas was adamant, and no doubt understood that it would have been difficult for men with such a roughneck, swashbuckling demeanor as drillers to turn down a woman whose request was nothing less than a challenge. The crew chief consented. A helicopter came in and slung a heavy drill out to the A-21. The ice conditions were worsening even as they finished setting up the rig. While water slapped against the drill shack, they sent the big drill down through the ice and water and into the pipe and began to harvest kimberlite.

As soon as the helicopter finished slinging gear onto A-21, it ferried the pieces of the lighter drill onto the new target, A-154. Rio Tinto had not liked this target at all, and did not favor drilling it. But Thomas had control of the Aber program. The airborne geophysics suggested a twinned body. With the water deepening on the surface of the ice, and the last hours of the winter drilling season melting swiftly away, Thomas and Hopkins decided to try for a hole into the center of the anomalies. They were desperate to make a strike here, too, because Aber's drilling budget was now stretched to the limit. If anything would support the results from A-21, it would be two more pipes at once.

They missed. The drill went straight down between the two bodies and, after two straight days, hit granite. Thomas faced the decision of whether to try a second hole. By this time the drill shack was awash. The pool around it would not drain away. Water was knee deep around the drill and the ice was dangerously soft under the surface water. The foreman told Thomas that he didn't like it, and thought they should pull the drill and get onshore. Another consideration was Rio Tinto. As senior partner, Rio Tinto bore most of the costs of exploration. It had protocols to govern drilling, and one of these banned the drilling of a second hole on a target where the first hole failed. There were dozens of targets to explore, and in the Barrens a single hole could cost C$20,000. But Thomas sited a second hole to the south. The drillers reluctantly agreed, and sloshed off, in growing consternation, to set up the rig again.

The drill cleared the surface gravels quickly and got in among boulders. It stalled, and the casing bent. The drillers hauled it up, replaced the casing, sent it down again. Black smoke puffed in the air as the drill strove to force a passage. The first shift passed. The late May sun blazed above them. Robin Hopkins remembers flying above the drills and noting how the weight was pulling down the ice. It was as if the whole surface of the lake were being slowly and inexorably dragged into an oily depression made by the heavy machines. Open water now appeared along the shore, creating a bank of fog that rolled out and obscured the rigs. When the shift change came around again, the fog was so thick the helicopter could not land at A-154. The drillers labored on in the rising water for twenty-four hours straight until the fog lifted and the helicopter came out with the relief.

Hopkins came out with the shift change. He waded over to inspect the core boxes, stacked up high on timbers that kept them above the ponded water. He peered at the first box, and it was full of kimberlite. Then he saw that the kimberlite was studded with indicators. The next box had even more indicators in the core, and the next one after that still more. "I kept taking samples and filling my pockets," Hopkins said. When he got back to camp he went straight to the kitchen, where he found Thomas sipping a cup of tea. She had a glum expression. She knew the drills had to come off the ice that day and had steeled herself to call it quits. Hopkins walked up to where she was sitting. "I started dumping core out from under my work shirt piece by piece. With each piece Eira's face got brighter and brighter until she was practically bouncing around."

As the drillers rushed to tear down the rigs and get them off the ice, Thomas called her father. The news of the strike on A-154 brought Bob Hindson, an Aber vice president, and John Stephenson, Rio Tinto's senior executive in Canada, into camp to inspect the core. With Hopkins and Thomas and two Rio Tinto geologists, they closeted themselves in the core tent and started picking through the core. The mood was high, and they traded jokes back and forth about the size of the diamonds they would certainly find. Then Hopkins, studying the broken end of a piece of core, discovered a particularly neat, triangular depression. He immediately searched for the matching piece of core. When he located it he found that the broken ends meshed perfectly together, and there, on the second piece, was the protruding crystal that had made the depression. It looked like the Muscovite micas they'd found before. Hopkins tried to scratch the crystal with his thumbnail, and couldn't. When he looked up, he saw that Thomas was raptly watching him. "No way," he breathed. He passed the core with the crystal embedded in it to Buddy

Eira Thomas in 1999, in front of the shaft that tunneled out under Lac de Gras to sample her discovery—the A-154 pipe. (Matthew Hart)

Doyle of Rio Tinto. Doyle stared at it. “No fucking way,” he said, and handed it on to Thomas. She took the core in her hands and gazed at it, a 2-carat diamond bouncing light from its crystal face.

The tent erupted. “We can’t tell anyone about this!” Stephenson bellowed in the canvas tent, broadcasting his jubilation to the whole camp. The explorers were exultant. Finding a visible diamond in a core is very rare. No one in the room had even heard of a diamond visible in a length of rock pulled straight from a hole, let alone a diamond weighing 2 carats. Stephenson insisted they seal the camp. No one would be allowed to enter or leave without clearance. Bob Hindson ordered the locks changed on the core tent and computer room. As a final measure they cut off the phone. Thomas wanted to call her father first, but Stephenson would not hear of it; the call could be intercepted. Thomas decided to physically carry the core to Vancouver to show her father. Stephenson opposed that, too, but Thomas set her lips in a tight line and told him she was taking it.

That night she slept with the core beneath her pillow. She flew to Yellowknife in the morning with Bob Hindson, caught the flight to Calgary, and connected to Vancouver. On the Vancouver leg she called her father, asking him to meet her at the airport. Gren Thomas complained that he had a cold, and would see her later. Eira Thomas persisted, and finally Gren agreed to meet Eira and Hindson at Hindson's house. When he arrived his daughter took him by the arm and led him inside. She sat him on the sofa and got him a glass of brandy. Then she took the length of core from her backpack and placed it in his lap. She handed him a lens and pointed to the diamond. Thomas touched the stone with his finger, and looked at it through the lens. Then he looked up at his daughter. She wore a smile from ear to ear. Thomas looked down at the core again.

"No," he said.

And Eira Thomas said, "Yes."

◆

Eira Thomas had discovered the highest-grade cluster of diamond pipes in the world. While the nearby pipe at Tli Kwi Cho did not produce the results its owners had hoped for, and DHK more or less collapsed, the pipes that Thomas found under Lac de Gras contain some 138 million carats of diamonds. The deposit will support a mine for twenty years and supply the market with an annual $400 million worth of rough.

6
The End of the Old Cartel

The richness of the pipes under Lac de Gras increased the threat to the De Beers diamond cartel. An important diamond lode had appeared in a country where the cartel held no sway. Deep drilling of the pipes outlined a huge resource; in a matter of years the Barrens would be producing some 15 percent by value of the world's annual supply of gem diamonds. Such an amount, marketed independently, would undermine De Beers's authority by hampering its ability to manipulate the diamond price. A more aggressive possibility was also avidly discussed—that Rio Tinto and BHP might form a cartel of their own, dropping their high-end goods straight down into the U.S. market, the world's largest.

Such conjecture is the daily bread of the diamond world, where the most baroque rumor will be seized upon and examined. The diamond cartel is not only a commercial structure but a way of thinking about a commodity. To those who accept its rationale, the cartel is the despotic ruler without whose hand the imperial borders would dissolve, and the Tartars would come charging in waving their terrible sacks of ungoverned rough, spilling it into the streets at bargain prices. "I tell you plainly," an Antwerp diamantaire once said, putting aside a parcel of rough and tapping his tweezers on his blotter to mark his words, "I hate De Beers. If you want to know, I hate them. But if they are not in business, if there is suddenly no De Beers, then you can have this whole office and everything in it." He stood up and swept his hand around, his face dark with anger. People stopped working and looked at him. "Because if they are gone, it is worth *nothing*! The whole place, all the goods — *nothing*! Go open the safe and take it all away!"

The man's wrath was no doubt partly a product of his helplessness. Like other customers of De Beers, he endured the iron grip in which the company, as paramount supplier of rough, necessarily had its customers. The emotions raised by this serfdom have given the cartel an imagined as well as a real power. Its real power is great enough: most of its customers would cease to exist if De Beers cut them off. So pervasive is this real power that it makes the company loom larger than life in the imagination of the trade, so that to conceive of diamonds without De Beers is to think of sailing off the edge of the known Earth. Like the Antwerp diamantaire, many look upon the Oppenheimers and their court as upon an absolute monarchy that nonetheless has kept the borders safe.

In South Africa, the idea of De Beers and the diamond cartel has had a similar hold on the imagination. The structure created by Sir Ernest Oppenheimer, in which Anglo American and De Beers and the Oppenheimer family were all woven into an impenetrable mass of crossholdings, sat at the center of the national economy and dominated it. When the apartheid regime collapsed and the African National Congress (ANC) came to power in 1994, it was natural to wonder what might happen to this massive block of capital that drove so much of the economy.

Tokyo Sexwale. (Trans Hex Group)

Would the ANC's program of black economic empowerment loosen the diamond stranglehold? Such was the background speculation when a powerful and charismatic black man made a move into the diamond world.

Tokyo Sexwale (seh-HWA-luh) was a veteran of the war for black emancipation. In 1975, at the age of twenty-two, he went to the Soviet Union to train as a guerrilla. He returned to South Africa the next year, where he was soon captured by police. Sexwale spent thirteen years of his imprisonment on Robben Island, the infamous political prison in Cape Town's Table Bay. His cell faced Nelson Mandela's across a narrow corridor. In the 1994 elections that followed the disintegration of the white government, Sexwale swept into office with the ANC. He became premier of Gauteng, the province that includes Johannesburg and Pretoria and is the locus of South African political and economic power. Sexwale had a fearless, flamboyant style; he once waded into an angry crowd of blacks to rescue a white policeman.

In 1998 Sexwale tired of politics. He judged that the tough fight for black advancement now lay elsewhere—in business. Of the total listed equity on the Johannesburg Stock Exchange, blacks owned a mere 6.5 percent. Sexwale thought even that figure was high, because some of the holdings were jointly owned with whites and, if properly diluted, would probably amount to no more than 2 percent. "Two percent ownership," he exclaimed, "for ninety percent of the population!" Sexwale resigned the premiership to enter business.

When he announced his decision, the white business establishment showered him with offers, including lucrative directorships that required nothing from him but his name. He turned them all down. He closeted himself in his house in the Johannesburg suburb of Houghton, and studied the patterns emerging in black empowerment commerce. Blacks had moved quickly into the media and information technology. They had also moved into the mining sector, taking stock positions in big companies through the use of advantageous credit arrangements available in the informal, but clearly understood, empowerment protocols. Sexwale noticed one large gap in this activity—diamonds.

In Sexwale's view, diamonds had a special place not only in South African commerce but in the country's mind. They had a mystique. In South Africa that mystique was firmly attached to the idea of power. At the head of the diamond world sat the Oppenheimer family, the richest family in South Africa. Diamond kings, they ruled a diamond kingdom. That was where the power lay, and Sexwale wanted power.

He moved swiftly, and within a year of his departure from politics he had targeted a group of alluvial properties along the middle reaches of the Orange River. With another black businessman, Wiseman Nkuhlu, Sexwale approached the owners, a successor company of the mining house founded by Barney Barnato. Sexwale knew the company wanted to sell off its diamond holdings and concentrate on gold. Sexwale and Nkhulu arranged the financing and took control of the diamond property; Sexwale became chairman. The concessions he then controlled were rich in high-end goods but small in extent, and the next objective was expansion.

Sexwale bought a peach-colored villa in the Johannesburg suburb of Rosebank and established the offices of Mvelaphanda Diamonds. Gardens surround the house. A purple Daimler gleams richly against the immaculate, raked gravel of the little parking lot. The lawn is as smooth as a putting green. Inside, the rooms are unusually spacious. The wide mahogany planks of the floor are polished to a high gloss. Sexwale's enormous office takes up half the top floor of the mansion.

In late 1998 he hired a Cape Town investment banker and laid down a strategy to enlarge his diamond position. He settled on four possible partners: Namibian Diamond Corporation (Namco), Ocean Diamond Mining (ODM), Benguela Concessions (Benco), and Trans Hex Group. Taken altogether, the companies had enormous ocean licenses off the Namibian and South African coasts. ODM and Namco were making steady profits from the seabed, and Trans Hex was taking high-value goods from its alluvial deposits on the Orange River. Of the four miners, the most substantial was Trans Hex. Although tiny compared to De Beers, Trans Hex was the second-largest diamond producer in South Africa, and the quality of its rough attracted eager buyers from everywhere in the world to its regular Cape Town diamond sales. Also important to a consideration of Trans Hex was who controlled it—the Rupert family of Stellenbosch.

Next to the Oppenheimers, the Ruperts were the richest family in South Africa. Like Nicky Oppenheimer, the present chairman of De Beers, Johan Rupert was a billionaire. The Ruperts had interests in banking, insurance, tobacco, and luxury products. Their companies owned such glittering brands as Cartier, Dunhill, Mont Blanc, and Lagerfeld. Through their Rembrandt Group, the Ruperts held 51 percent of Trans Hex. Other large blocks of the diamond miner's stock rested in the hands of institutions with long-term investment strategies. With such large stock holdings more or less immobilized, the stock was sluggish. Sexwale believed that the market was undervaluing Trans Hex, that the shares were worth more than they were trading for.

Sexwale's advisers studied Trans Hex. With the help of a Johannesburg merchant bank, they structured a deal in which Sexwale would exchange his Orange River mines for shares in Trans Hex. Trans Hex's richest mine was also on the Orange

River, and the Cape Town company had expertise both in mining and marketing such high-value alluvials. In early 1999, when everything was in place, Sexwale flew to Cape Town. A car met him at the airport, and he drove to Stellenbosch to meet Johan Rupert.

The irony of that meeting must have been apparent to all. Sexwale embodied the new force of black empowerment, and yet the term "empowerment" had also been used to describe the push of Afrikaners in the 1950s and 1960s into a business world dominated by men of English descent. Afrikaner empowerment had been nurtured by the white nationalist government, the same government that introduced apartheid. "These people's lawyers wrote the apartheid laws," Sexwale's chief aide said of the Stellenbosch businessmen, "and here they were meeting someone who destroyed those laws. And Tokyo liked them and they liked him."

On May 15, 2000, Sexwale acquired 8 percent of Trans Hex and became deputy chairman. Insiders close to Trans Hex said Sexwale's holdings in the company would eventually increase to 25 percent. In one move, Sexwale made himself an important diamond player. Trans Hex got the diamonds of the middle Orange River. Johan Rupert, in exchange for a minority piece of a small possession, gained the distinction of helping a prominent black advance into the tight world of diamonds. In the game of appearances, the Ruperts had scored on the Oppenheimers.

Sexwale's entry into diamonds, though relatively modest, was a development with considerable allusive force, a symbol of the new South Africa chipping at the old. It seemed to fit the spirit of the times, a spirit hostile to the old cartel. By the time Tokyo Sexwale took his new place, the borders of the middle empire were already gone, and one of the forces that had breached them had come from Canada.

◆

An important force unleashed at the time of the diamond rush in the Barrens was speculative capital—a pool of new investment created by the rapid enrichment of a host of junior exploration companies. Awash in money from an overheated market, the juniors carried the diamond hunt to many corners of the world. Exploration heated up in Brazil and Uruguay. Prospectors examined targets in Russia, the Ukraine, and Finland. Robert Gannicott found kimberlitic boulders in Greenland. There was a diamond discovery in Colorado, and in Canada the exploration spilled over into Alberta, Saskatchewan, and Ontario. Juniors made discoveries; seniors came in behind. Indeed, De Beers's first diamond "success" in Canada came by buying out the junior Winspear Diamonds, in a brief but spirited war of words and stock market bids that captured the Snap Lake diamond dyke for the cartel. The outward rush of diamond prospectors into the world was an exhilarating and, for De Beers, alarming outcome of the diamond strike. But few diamond hunters were as brazen as Chris Jennings, who took the diamond hunt back to where it had come from, South Africa. Jennings reasoned that advances in prospecting might help locate deposits missed in the past, especially by De Beers, and he went to work in the diamond titan's front yard.

In 1994 Jennings met Hennie van der Westhuizen, a retired De Beers geologist. Van der Westhuizen had discovered a kimberlite fissure in the hills above a plain called Springbok Flats. A fissure deposit is much smaller than a pipe, and De Beers had not pursued it. But the geologist thought it was worth a closer look, and when Jennings heard about it he agreed. He secured the mineral rights, and a team went in to dig an exploration trench. They found that the fissure ran for several miles through the hills. Then they found a parallel fissure, and then a "blow," where the fissure widened into a sort of micro pipe.

By 1996, Jennings's SouthernEra Resources had acquired 100,000 acres of mineral rights. Workers sank a shaft and began the laborious process of extracting diamonds. Geologists tracked the main fissure until they had mapped a system that ran twenty-five miles through the choppy hills.

A mood of elation took hold at the exploration camp. Jennings rented a massive old farmhouse in a grove below the hills. Purple and crimson bougainvillea bloomed beside the house. Frangipani dripped their perfume into the air. Every evening young geologists gathered at the farmhouse and discussed the amazing fissures. At night on the lawn they barbecued *boerewors*—South African country sausage—and planned the next day's work. Next morning at six o'clock they piled into pick-ups and bounced back into the stony hills. Then, on a farm called Marsfontein, they found a pipe.

The discovery of the small, rich Marsfontein diamond pipe sent SouthernEra's share price climbing. The company had other successes too. In Angola it won the right to explore the Camafuca kimberlite pipe, one of the largest diamond-bearing pipes in the world. Under the deal, SouthernEra would pay for exploration and share the profits with the Angolan government. The Camafuca deal was a coup for SouthernEra, and a leading diamond commentator called it a bitter blow for De Beers. With the Camafuca deal complete and the news from Marsfontein billowing in its sails, SouthernEra's share price reached the record high of C$20. On the diamond street, people were beginning to call De Beers a dinosaur. The plodding cartel had been unable to prevent the quick-footed junior from swiping the Marsfontein diamond pipe out from under its nose, two hundred miles from De Beers's head office. At the farmhouse in the frangipani trees, things had a wonderful, fairy-tale feel. Then everything came apart.

Late in 1997, heirs to the former owners of the Marsfontein farm came forward to contest SouthernEra's right to the minerals. SouthernEra responded in court. The fight dragged on for months, and SouthernEra's share price started slipping. One leading Johannesburg analyst downgraded the stock from a "buy" recommendation to a "hold." The share price went to C$12.00. The worst was not over.

On April 14 SouthernEra's lawyers arrived at the High Court in Pretoria for a scheduled hearing, and learned that the heirs had sold their rights to De Beers. Without warning, SouthernEra found itself facing the dinosaur on its home ground, with legions of lawyers at its beck and call. Investors took one look and began to dump SouthernEra stock. In one bloody week, the market hacked a third of the value from the shares.

As the legal wrangling wore on, it took its toll on the sixty-four-year-old Jennings. His face became drawn. In the ordeal with De Beers, Jennings and his wife, Jeanne, were together all the time, poring over legal documents, sharing each other's bitterness. When they could, they left Johannesburg for the refuge of the farmhouse beneath the hills, but the carefree mood was gone. At night the Jenningses' voices could be heard from their rooms on the ground floor, where they stayed up late and worked, surrounded by stacks of legal briefs.

Jennings finally bowed to the inevitable. De Beers took 60 percent of the Marsfontein deposit. Another half million carats a year joined the tens of millions flowing to De Beers. The dinosaur had stamped its foot—the last spasm of the old cartel. After Marsfontein, different preoccupations took over at De Beers, so that gobbling up each stray carat no longer had priority. Instead, the company would fashion a stronger instrument from the grand stream of diamonds it already controlled.

◆

The true capital of the diamond world lies on the northwest edge of London's financial district, in Charterhouse Street, at Number 2 and Number 17. No sign identifies the buildings. Number 2 is a plain gray block, with diamond sorters' flex lamps visible in the windows. Number 17 is newer and somewhat uglier. Nothing gives the passerby the slightest clue to what is inside. The tinted windows of Number 17 are suitably opaque. From the men in pinstripe suits going in and out you might think it was a business like any other instead of an enterprise where the arcane is not simply a practice but the product.

A torrent of diamonds pours into Charterhouse Street, hundreds of thousands of carats a week, tens of millions of carats a year. Some 60 percent of the world's production of rough dia-

monds (it used to be 80 percent) arrives in armored trucks by night and day at the Diamond Trading Company (DTC), a De Beers firm that sells the rough produced by De Beers as well as the rough it buys from other sources, such as Russia and Canada. The DTC sells rough at ten sales a year, called "sights." A buyer may not simply show up at a sight and hope to buy diamonds. He must be a "sightholder," which is to say a diamond trader or manufacturer with standing in the trade, good finances, and, most important, an invitation from De Beers to become a sightholder. Only this elect group may come to Charterhouse Street. The rest must rake about in Antwerp and Tel Aviv and Bombay to buy loose rough sold outside the De Beers cartel, or rough resold by those who have bought it from the DTC. The sights are the most important diamond sales in the world. Through them, De Beers directs hundreds of millions of dollars' worth of rough into the cutting centers, every five weeks. The diamond world thus revolves around a five-week cycle, from sight to sight. While the customers are off polishing the goods they have just bought, the DTC gets ready for the next sale.

The DTC is an organism of De Beers, and is the successor of the original Diamond Corporation established by Sir Ernest Oppenheimer to replace the syndicate of London dealers. Sir Ernest's Diamond Corporation gave way to the DTC, which was owned not only by De Beers but also by Anamint, an investment trust of Anglo American. The Oppenheimer family also had a direct stake in the DTC, as well as whatever they owned by virtue of their shares in De Beers and Anglo. As of this writing, a new arrangement was being proposed, whereby the Oppenheimers, Anglo, and a smaller partner would buy out all the shareholders of De Beers, creating a private company removed from scrutiny and sweeping from view all that was known of De Beers and its entanglements. Presumably, though, even such a secretive entity would still need someone to sell its diamonds, and that would be the DTC.

The whole thing is a carnival of appellations. Some Antwerp traders still call the London setup "the syndicate," harking back to the good old days, in the distant pre–Sir Ernest past, when the syndicate called the shots. Other sightholders use the later designation, CSO, for Central Selling Organization, which was for thirty years the De Beers term for the London operation, until the company suddenly stopped calling it the CSO and started using DTC. I asked Nicky Oppenheimer, Sir Ernest's grandson, about the status of the CSO, since I still had calling cards from some of his employees that bore this acronym, and since the term was plainly in wide use. "The CSO?" he said, as if surprised to even hear such a term. "It doesn't really exist. It has no basis. The top echelon in London call themselves directors of the CSO, but it doesn't really mean anything."

Just so it's clear.

The five-week cycle begins when the tin boxes of rough arrive in the receiving room in Charterhouse Street. The boxes are unlocked and the contents check-weighed. Next, handlers dump the larger rough into the hoppers of the diamond-sorting

Sorters at the Diamond Trading Company, London. (De Beers)

machines, each about the size of a soft-drink dispenser. The machines are known by the acronym SADE—Scale Automatic Diamond Electronic. Rows of SADE machines chatter away on the fourth floor of 2 Charterhouse Street, chewing through bag after bag of rough. The goods tumble down chutes into plastic collection boxes at the bottom. The steady tick-tick-tick of falling diamonds sounds pleasantly in the room. The SADE system weighs, counts, and sorts by weight every piece of rough between half a carat and 10 carats. Through a computer link, it also calculates the value of the parcel. Larger goods, called "specials," are sorted by hand. Smaller sizes go through sieves—metal discs punctured by holes. The smallest sieve size separates out diamonds so small that it takes ninety of them to add up to 1 carat.

De Beers sorts rough into more than fourteen thousand categories, including the finest distinctions of crystal shape, size, color, and clarity. Many of these categories are unique to De Beers, and the sheer profusion can be exasperating. South Africa's government diamond valuator (GDV), Claude Nobels, an Antwerp diamantaire, seethed with resentment when he described arriving in Kimberley to value De Beers goods and finding the tables spread with fourteen thousand categories. "It takes them three floors at Harry Oppenheimer House to lay out all the goods!" Nobels fumed. The problem for Nobels was that, if he disputed the category in which De Beers had placed some rough, they might move it up into the next, higher category. But such a move—one step on a ladder with fourteen thousand rungs—resulted in only a tiny upward adjustment in the overall value. By contrast with this system, the GDV for Canada sorts the goods into only a few hundred categories, and extrapolates a market value for the whole production.

Uneasiness with the De Beers sorting regime has troubled even members of the cartel. Perhaps the most famous account of a dispute between De Beers and a diamond miner is that contained in *Burning Bright*, the autobiography of a mining executive named Edward Wharton-Tigar. In the 1950s, Wharton-Tigar ran diamond operations in West Africa for the British mining group Selection Trust. Sir Ernest Oppenheimer had brought Selection Trust into the fold of the cartel, and so the West African production went to the Diamond Corporation in London to be sorted and valued, and sold through the CSO (as it then still was). Wharton-Tigar doubted that the cartel system, in which the buyer of the goods was also the valuator, could be fair. As he described events, he sent 1,000 carats of his choice Sierra Leone production—some of the best rough in the world—into the Accra market, where it was promptly valued at twice the valuation put on a similar parcel by the Oppenheimers' Diamond Corporation. In June of 1956 Wharton-Tigar wrote to De Beers seeking an explanation.

In Wharton-Tigar's account, the effect was immediate. Philip Oppenheimer, Sir Ernest's nephew, flew into a rage, accusing the executive of meddling in affairs he did not understand. This broadside was quickly followed by an attack on Wharton-Tigar by one of his own directors, who happened also to occupy a similar post on the board of a De Beers company. The director told Wharton-Tigar that if he did not learn to get along with the Oppenheimers he would have to go. Wharton-Tigar laughed at this, and would not be budged. De Beers then wrapped its mailed fist in a velvet glove and entertained Wharton-Tigar at a series of what he called "smoked salmon lunches" at De Beers's executive dining room in London. Wharton-Tigar held to his position. At the time, vigorous smuggling of top Sierra Leone rough was flooding into Antwerp, and De Beers did not want to lose the whole of the production. A compromise was reached.

Nonetheless, Wharton-Tigar remained suspicious. A month later, according to his book, he forced De Beers to surrender the master sample of Selection Trust's goods, the marker against which production was supposed to be valued by the sorters of the Diamond Corporation. He secured an independent valuation of the sample, which showed that the values were out of date and should be raised. De Beers agreed, but Wharton-Tigar soon realized that no real revision of the price would follow, and he ordered the London deliveries stopped.

De Beers applied all the pressure it could on the upstart diamond miner, including a reprimand from an official of the Colonial Office. Wharton-Tigar was unmoved. Finally, in the summer of 1958, Harry Oppenheimer, Sir Ernest's son, invited Wharton-Tigar to Johannesburg to stay with the Oppenheimers at Brenthurst, their walled estate. Harry Oppenheimer was a hard and temperamental negotiator, and Wharton-Tigar has reported that the two men battled it out for days. In the end they agreed on a higher price for the West African goods, and Sierra Leone diamonds again moved to the CSO.

However, this was not the last of the affair. As Wharton-Tigar remembered it, although Selection Trust was getting higher prices for its diamonds, its shipments suddenly seemed to contain more industrial, low-price diamonds than before. The percentage of valuable "cuttables" in the overall production had decreased in a proportion that perfectly neutralized the price increase. Where London's sorters had been finding 20.4 percent of the goods to be cuttables, they now found only 17.8. Wharton-Tigar devised a test. He ordered 1,000 carats of the best gems to be removed from each of the next three shipments to London. The percentage of cuttables barely shifted.

From the next shipment, as his memoir has it, Wharton-Tigar removed *all* recognizable top gems. He then visited the London chairman of the De Beers subsidiary that sorted his goods, and asked him to pay particular attention to the gem count in this latest shipment. The De Beers man duly reported back a figure of 18.5 percent cuttables. Wharton-Tigar then took out a bottle of diamonds from his briefcase and placed it on the man's desk and asked him what he thought about it. The man dumped some out on his blotter and fingered through them and said, "These are the best cuttables from your Ghana goods." Wharton-Tigar agreed that indeed they were, adding with some satisfaction that the diamonds had in fact been removed from the very shipment in which the sorters had just reported finding them.

When Harry Oppenheimer heard of this, he professed astonishment. De Beers agreed to revalue all Selection Trust's shipments of the previous two years, and according to Wharton-Tigar, he received an adjustment check of £250,000.

◆

At the long benches that run beneath the north-facing windows of 2 Charterhouse Street, the sorters hunch over piles of diamonds arranged on spotless white paper. The rough is already clean, having been washed in acid at the mines, and again in Charterhouse Street, to remove the eons of grime. The sorters arrange the goods in tidy little mounds. The abundant natural daylight and the light from their lamps illuminate the piles of rough and the busy, ceaseless flash of tweezers and loupes as the sorters pick through the goods. Some 250 diamond sorters work in Charterhouse Street, and to see long rows of them bent above very different heaps of rough is to grasp the bewildering range of color, size, clarity, and shape into which rough is sorted in the greatest diamond house on Earth.

There are small piles of brownish goods, some that are only barely whiter, and some that blaze with white radiance, like heaps of shredded ice lit by some powerful interior source. There are large goods and small, spotted and clear, wonderful, lemony Cape goods, sometimes a few piles of them in a row, each a little less yellow than the last. The finest gradations of color become apparent when such large volumes of diamonds are sorted into categories and laid out along the benches. Sometimes a sorter will pluck a stone with his tweezers and toss it onto a scale with an audible "plink," retrieve it and drop it in a pile. There may be a hundred different groupings of rough heaped along the sorting bench at one time, with tweezers picking through it. With all these diamonds laid out in such a rich display, the mind turns irresistibly to the question of how to steal them, because people do.

At every stage of the pipeline from mine to consumer, someone is trying to steal diamonds. In the sorting rooms of Africa, the pockets of the sorters' shirts are removed, and their pants pockets sewn shut. In London there is no such practice, and yet there are more goods to steal. A popular form of theft is "switching." A sorter may replace a higher-value stone with a lower, which he has brought to work. Switching is endemic to the trade. It takes advantage of a universal practice—the control of diamond inventories by weight.

At a diamantaire's office in Antwerp, for example, parcels provided to clients for inspection are first check-weighed against the figure penciled on the packet, then handed over. There are cameras on the ceiling, as there are in Charterhouse Street. But cameras often fail to catch a talented switcher, who keeps his back to the lens and whose movements are deft. So the switcher flicks a diamond into his cuff and replaces it with a cheaper one of the same weight. Out comes a half-carat top color, in goes a half-carat of a lesser color. The packet is handed back, weighed, and because the weight tallies, put back in the safe. The story is told of an Antwerp trader who had a tiny rubber tube in his sleeve. It attached to a rubber bulb under his armpit. First he would expel the air from the line by squeezing the bulb with his arm, then place his cuff near the stone he wanted and raise his arm. Air whooshed back into the line and so did the diamond.

At the DTC, a volume of goods found nowhere else flows from department to department. If a sorter or diamond handler comes to work with a small stone, he may steadily "switch up," gradually increasing the size of the switched stone. Each time he switches, only a fraction of weight would be lost from the stream, a weight loss undetectable within the margin-of-error assumptions of the DTC. In the end, an employee who brought in a quarter carat might leave with 5 carats. But this is not the usual way to steal from the DTC. The usual way is just to take a diamond, and to hell with switching.

In the case of plain theft, an employee relies on the huge volumes of the DTC to conceal his peculations. In the early 1990s, Charles Wyndham, then chief of security at the DTC (and now a principal of WWW International Diamond Consultants) decided to see what he could do about a situation he had inferred. He did not know for certain that diamonds were being stolen in Charterhouse Street, but suspected they were. "There was this attitude that people stole diamonds everywhere else, but not in London," Wyndham said. "I thought, why wouldn't they steal diamonds there? There are more diamonds to steal."

In the hypothesized London thefts, Wyndham identified the large mass of diamonds as the field of the problem, and a certain industry practice as an accident that tended to favor theft. The practice was that of rounding down weights. The DTC weighed diamonds to a degree of accuracy of several decimal places. For practical reasons, such as the volume of goods, they rounded down the weight of parcels to two decimal places. A carat weighs only .007 ounces, so recording weights to a fineness of two decimal places is quite exact. The decision to round down to two decimal places reflects a decision that the accuracy of a third decimal place is excessive. Nevertheless, the third decimal place is there, whether one chooses to express it or not. If you have a weight of .282 carats, for example, and round it down to .28, then a weight of .002 carats has gone missing. This is not hypothetical weight, but real weight, although very small. It is, in effect, weight that has been stricken from the parcel.

In practice, you cannot steal .002 carat of diamond, because it is too small. But if you understand that the .002 carat has disappeared, you may safely skim off a stone here or there, secure in the knowledge that in the huge volumes of goods pouring through the DTC, the fractions of lost weight, steadily accumulating, will eventually equal the weight of whatever you have stolen. Wyndham knew this, just as he knew that over time the pilferage could add up to a substantial loss.

He began to monitor the diamond flow more intensely. He and his staff would scrutinize the long lists of totals, searching for the tiniest discrepancy. "This threw up anomalies," says Wyndham, "which in themselves were not surprising, given the incredible volumes of carats within such an overly complex sorting system, seventeen thousand price points [as it was then]. It is the pattern of these anomalies that was important." Whenever they found such a pattern, the whole force of the audit (and of the "special audits," as cameras are called) would be directed at the department where the anomalous pattern was found. Wyndham would then pick a departmental employee at random from the company phone book. The full scrutiny of the whole security staff would shift to that employee.

"The tapes from the security cameras need a lot of examination," Wyndham said. "Those watching the cameras received special courses in body language. You have to know what to look for. After all, the people know the cameras are there." Since handling and sorting diamonds requires the repetition of certain movements over and over again, Wyndham and his staff particularly watched for the slightest variation in the actions of the employee under surveillance. Any divergence from the customary pattern would trigger the watchers' keenest attention. With patience, and the consecutive focus of his full staff on one employee at a time, Wyndham caught thieves, although, as he said later, "no doubt it's still going on."

◆

The whole first week of the cycle is devoted to sorting diamonds. In the second week, the sorted rough passes to Diamond Control. At this point De Beers begins to re-sort the diamonds into selling mixtures. Different sightholders have different needs, since they have different downstream clients. But De Beers must sell all its diamonds, not just those its customers want. Diamond Control devises elaborate mixtures of goods, called selling mixtures, which contain both the preferred goods of the clients, and those they may not want. When the makeup of the mixtures for the next sight has been fixed, the DTC faxes out lists to its clients. The clients pick what they want and fax the list back to licensed brokers, who are the only ones allowed to negotiate with the DTC.

In the third week of the cycle De Beers prepares the clients' boxes. In former days the diamonds were put in gray cardboard boxes the size of a shoe box, and they are still shipped out that way. But the term "box" also means a client's order. As De Beers is making its final decisions about the box, the clients' brokers do their best to keep unwanted goods from the box and include only desired goods. Every client of the DTC must have one of these brokers, who are licensed by the DTC. There are only six of them. They are supposed to be at arm's length from the DTC, but in fact have strong ties to it. Their livelihood depends upon being acceptable to De Beers. One of the brokers, I. Hennig & Son, has its office in the De Beers building, suggesting that De Beers might actually own it. I put this to Nicky Oppenheimer. "No," he said, "that's not true."

"Perhaps the Oppenheimer family owns it?" I asked.

"It could indeed."

Oppenheimer is a bearded, amiable man, faultlessly polite, whose passions are his private cricket team and his helicopter, which he flies himself. We had just concluded an interview, and I was taking my leave in the hallway outside his old office in the

Anglo American building in downtown Johannesburg. He regarded me calmly as I pestered him about the diamond broker.

"In fact," I said, "it could be owned by you."

"Yes," Oppenheimer agreed with a sigh, "it could." And he shook my hand and walked away.

The assortments provided by De Beers for clients to choose from are complex. The DTC changes its assortments constantly, depending on the goods it wants to sell, but one example of a box that was a staple offering for years was MLG 2.5–4. The initials stand for Mixed Large Gem. Despite the numerals, the diamonds in MLG 2.5–4 actually weigh between 2.49 carats and 3.89 carats. The box would include stones, shapes, and cleavages. In DTC terms, "stone" means a desirable crystal shape, possibly a good octahedron, one that will transform into a polished gem with a minimum loss of weight. The next classification, "shape," means a weaker shape, one that will lose more of its weight to the polisher. Cleavage means pieces of broken rough that usually go straight onto the polishing wheel, and deliver a lower-than-average yield. The contents of the box may be top or spotted, meaning clear or included. (An inclusion is an impurity.) The color will range from white to brown, or first color to fifth color, which is how rough is graded. (Polished diamonds, in the system used by the Gemological Institute of America, are graded by letter, with D being the top white color.)

When the brokers have agreed on the box—as, in the end, they must—De Beers handlers put the diamonds into plastic bags and lock them in dark blue attaché cases trimmed with yellow. The cases go in a locked room. In the fourth week the DTC checks everything over. If the client's broker has managed to negotiate any last-minute changes to the box, those changes are made. The attaché cases are made ready for the clients.

Week five is sight week, the most important week in the diamond calendar. In the first-class cabins of jets from New York and Tel Aviv and Bombay, on the shuttle from Antwerp to London's City Airport, come the world's biggest diamond dealers. Some of them repair to $700-a-night hotel rooms on Park Lane and sag onto silk-covered sofas while their brokers tell them what they got and did not get. They may make a few calls to other diamantaires, or even go out to dinner, but mostly they get what rest they can.

On the appointed day, the sightholder gets out of his taxi in Charterhouse Street and hurries into Number 17, where a guard surveys him from behind thick glass. When the outer door has closed and locked behind the client, the inner door buzzes open. On the second floor, porters in dark jackets stand behind a wooden counter and check the client's name from their list. To the right of the counter are some plain sofas where the brokers wait. When his client arrives, the broker reports to a grille at the end of the reception area and collects the "box"—usually consisting of several attaché cases. The cases are loaded on a cart, and a porter trundles them off to whichever room has been assigned to the sightholder. The client has an idea of what is inside the cases, but not a certainty, making this surely the only transaction in the world where a man may arrive to spend $200 million without knowing what he is going to get.

The cavalcade of porter, broker, and client enter a plain room with gray vinyl walls and a blue carpet. A round table occupies the center of the room. Under the ample windows is a sorters' bench with a padded elbow rest. Sheets of white paper cover the surface of the bench. Flex lamps are poised above the surface. The porter leaves, and the broker and sightholder carry a Samsonite case to the bench, open it, and take out a bag of rough. They spill the contents onto the paper, take out their loupes, and begin the process of discovering the client's fate.

"We used to call it 'feeding the ducks,'" said Richard Wake-Walker, a former executive of the DTC. "The ducks come paddling over and you throw them the bread and they eat what you throw." Wake-Walker was once responsible for creating the selling mixtures that went into the boxes. He helped develop allocations policy, and is one of the highest-level employees ever to leave De Beers and speak openly about its business. It was a troubling desertion for De Beers, and Wake-Walker remembers a dinner with a senior member of the company after he'd left. "He'd invited my wife and me. He began by offering me a lucrative consulting job. I said no, that wasn't my plan. Then, in the presence of my wife, he said De Beers could do certain things to make things hard for me, very hard. It made me very angry."

Wake-Walker would not change his mind, and he and Charles Wyndham set up WWW International Diamond Consultants, which numbers among its clients the Canadian and Russian governments, as well as numerous manufacturers and diamond miners. He runs the consultancy from a small, bright office at the front of his house in the London suburb of Wimbledon. A drawing of his great-great-grandfather, a naval hero, hangs in the hall. Another naval ancestor, his grandfather, Admiral Sir Frederic Wake-Walker, commanded the flotilla of warships and fishing vessels that sailed across the English Channel to rescue the British army from Dunkirk on the French coast, and later in the war was instrumental in the pursuit and ultimate destruction of the German battleship *Bismarck*.

Richard Wake-Walker went into the diamond business because it promised adventure. He joined the DTC in 1975 at the age of twenty-four. After an intensive course in sorting and valuing rough, he went to Sierra Leone as assistant resident manager of the Diamond Corporation of West Africa, the De Beers subsidiary in the country. One year later he was in Kinshasa, Zaire (now the Democratic Republic of the Congo), where he became head of the De Beers subsidiary, with a staff of 750.

After a stint in Johannesburg attached to Harry Oppenheimer's staff, Wake-Walker returned to the DTC in London, where he moved swiftly into the top echelon. He became secretary of the executive and management committees. Wake-Walker knows what De Beers clients must endure because he helped inflict it. "All that happens is that they sit in their little room and open their box of diamonds and look at it. If they want to talk to anyone about it, they sometimes have to sit around for hours and hours waiting for someone from the sales department. They were *made* to wait, and sometimes they missed flights."

Wake-Walker's knowledge extended into the highest councils of the cartel. He ran the first De Beers office in Russia. In London he ran the West African department, and had such key jobs as client selection and, more chilling, "deselection." Because of his position, Wake-Walker says, clients regularly deceived him. "It was quite a revelation [to leave De Beers]. One of the things I found right away was how differently people spoke to me. I realized I'd been strung a load of bull for eight years. When I was at the DTC, no sightholder would ever admit that they'd made large profits, because that would have meant I'd put the price up on their box the next time around."

De Beers pays close attention to what its sightholders are getting for their goods. The company has several windows into the trade. Five of its sightholders, for example, are in fact owned by De Beers, including Diamdel NV, in Antwerp, Hindustan Diamond Company, in Bombay, and similar rough-diamond traders in Tel Aviv, Johannesburg, and London, making De Beers one of its own largest customers. It has as well a polishing department in London, which takes boxes identical to those being sold and polishes the goods, providing the sellers of rough with ready proof, close to hand, of the kind of jewels, and therefore the prices, that their customers can extract from the goods. Even independent sightholders can be useful sources of business information, and some are said to curry favor with De Beers by reporting the activities of their competitors. Sightholders favored by De Beers can earn bonuses. "We didn't hand out envelopes stuffed with cash," Wake-Walker said, "but there was a Christmas allocation of specials, and they went to valued clients. We might charge them twenty thousand dollars for a stone worth, say, two hundred thousand dollars."

◆

De Beers ran its cartel by holding rough back from the market when the price was low, and releasing it again when the market had sold off its stocks and was desperate for the goods. Mines controlled by the cartel were put on production quotas. When there were too many diamonds around and the price began to wobble, the cartel slashed output and raked loose goods out of the Antwerp market, stockpiling them in London. By the mid-1990s, the London stockpile was valued by De Beers at about $4 billion. This valuation was questionable, both because De Beers itself had set it, and because any attempt to rapidly convert the stockpile into cash would have collapsed the diamond price.

Financial analysts who followed De Beers discounted the value of the stockpile when they assessed the company. As one writer put it, the diamonds were "accumulating dust instead of interest." Moreover, De Beers earned most of its income from its shares in Anglo American, and significantly less from selling diamonds. One might conclude, and people did, that De Beers could not run the corner store.

In 1998 De Beers retained a Boston consulting firm to conduct a management review, the first independent review of De Beers in its history. The consultants advised the company to abandon the role of "market custodian," and let the diamond price take care of itself. By hoarding goods and cutting back production at its own mines, the consultants said, De Beers was subsidizing its competitors, such as the new mines in Canada, who would enjoy a free ride on the stable price supplied by De Beers through the self-limiting measures of the cartel.

The advice to get rid of such notional market supports as the stockpile came at a time of unmatched prosperity in the United States, the biggest diamond market. In response to strong demand, De Beers shoveled its stockpile out of Charterhouse Street as if it were clearing the driveway of snow. In the first half of 2000, compared to the same period the year before, De Beers profits rose by 226 percent, to $877 million. Writers in the diamond press heralded the "postcartel" era, an appellation encouraged by De Beers but somewhat meaningless, since the company still controlled almost two-thirds of the world's rough.

As De Beers swings its strategy away from the century-old doctrine of an absolute control of rough, the downstream diamond market—jewelry stores—will feel the change. In early

2001 De Beers announced with much fanfare a joint venture with the French luxury retailer LVMH, the proprietor of such high-price brands as Moët & Chandon champagne, Louis Vuitton luggage, and TAG Heuer watches. The partners will spend $400 million over five years to create a new chain of luxury stores devoted to the sale of De Beers brand diamond jewelry. Rough diamond sales from mines are worth about $6 billion a year, but in the same period the retail trade rings up sales of some $56 billion. This figure is expected to grow by 2.5 percent a year over the next ten years. De Beers has long coveted a share of the lucrative retail trade, and as its former control of the rough market erodes, income from its new stores will flow in—$500 million a year at the optimistic end of the projections.

In this new diamond universe De Beers will enjoy a ready-made advantage over other sellers of diamond jewelry, for De Beers ads are already famous. The new merchants of De Beers jewelry will step straight into the glitter of the greatest diamond name in history. It is a neat trick. When the De Beers stores open, De Beers's retail competitors will be forced to add their own diamond advertising to the retail scene, thus stimulating the downstream appetite for the product that, alone among retailers, De Beers also sells—millions upon millions of carats of rough.

7 The Manufacture of Desire

In 1997 a prominent Antwerp diamantaire stepped to the microphone at a diamond conference and told an audience of miners, prospectors, and financiers that the entire diamond business rested on two supports—vanity and greed. Fortunately, he said, the human race could be relied on for a perpetual supply of both. This sunny assertion is a favorite of the diamond trade, but in fact it says little about how things work. De Beers sells $500 million worth of diamonds every five weeks to the same people, and those people sell them in turn, and after them, someone else. By the time the diamonds reach the jewelry store their price is ten times what De Beers sold them for. To rely upon vanity and greed exclusively to speed the sale of these goods would be to believe that such vices were the whole of the human condition, and De Beers does not believe it. De Beers believes in advertising.

Diamonds are sublimely useless. You cannot eat them or drive them home. In the common sizes they are unreliable investments. Tradition, certainly, supports a diamond's value, but with hundreds of millions of individual diamond stones coming out of the ground each year, tradition is not enough. There have to be more reasons to buy diamonds, and there are: De Beers has invented them. The company spends $200 million a year advertising diamonds, and it does the job so well that the advertising professionals judge De Beers ads to be among the most successful in history.

The company's millennium campaign, for example, helped create demand that increased rough diamond sales in the first half of the year 2000 by 44 percent from the same period the year before. Sales for the period went from $2.4 billion to $3.5 billion. The message that helped to accomplish this was a classic diamond pitch. "Show her you'll love her for the next thousand years," the millennium campaign exhorted, displaying with perfect economy two of the assumptions that drive the marketing of diamonds: that men buy them, and that they buy them for women.

Statistics support this. Eighty-five percent of American women own at least one piece of diamond jewelry. In the case of married women, 80 percent of the diamond owners received their diamond as a gift, usually from a man. With single women, 64 percent received their diamond as a gift, and again, usually from a man. A decadeslong campaign of advertising has made sure the numbers fit the assumptions, so that it now seems like an inescapable archetype that men give diamonds to women. It was not always so. Louis XIV did not buy diamonds solely for his queen or mistress; he wore most of them himself. If the conqueror Nadir Shah had a girlfriend, she had to wear something other than the Koh-i-Noor, which was for him. But today there is a limited supply of monarchs and a large supply of diamonds. Every day of the year another 328,000 carats come out of the ground, and someone must buy them. The ingeniousness of De Beers's marketers lies in having forged a link between something people do not need, diamonds, and something they do need, love.

In the late 1930s, the demand for diamond engagement rings was falling in the United States, even then the world's chief market for diamond jewelry. Sir Ernest Oppenheimer hired the New York agency N. W. Ayer & Son to create an appetite for engagement rings. Men were the earners of money, the agency reasoned, and would have to be the givers of diamonds. The company established its campaign on an imaginative terrain that would prove fruitful for diamonds—eternity.

> [1939] There are many things a man must consider when undertaking one of his lifetime's most important purchases—his diamond engagement ring. . . . That with this symbol he institutes a new dynasty which will bear his name beyond his generation.

> [1940] No other treasure of earth or sea which they may acquire in later life will ever have one-half such precious significance for them. That is why it must be chosen as if already the weight of years and dignity overshadowed them.

> [1941] For in his heart each is convinced that of all the curly-headed girl babies born her year, this one was destined to be his as surely as the chubby sons and gurgling little daughters he hopes some day to hold. That is why each man gives, at his engagement, a diamond ring. For one day, close to the birth of time, his diamond, too, was set aside to make his happiness.

Every year the copywriters stretched for another phrase. In 1946 they invoked a star that "died a million years ago. It shines anew bright for her because she dreams of love." And then one night in 1948 an overworked, exhausted copywriter came up with the line that put all others in the shade. Frances Gerety of

N. W. Ayer had been working late on a De Beers ad for presentation to the client the next day. She had finished her work and was putting things away when she realized she had forgotten the signature line. "Dog-tired, I put my head down and said, 'Please God, send me a line,'" Gerety recalled. Then she sat up and wrote "A diamond is forever."

Within three years of Gerety's late-night inspiration, 80 percent of American marriages were starting with a diamond ring. The trade magazine *Advertising Age* has called Gerety's four-word line the greatest advertising slogan of the twentieth century. The phrase "a diamond is forever" has passed into idiom, as if it had grown out of some inevitable consensus instead of from a copywriter's pen. To make sure no one forgets that the slogan does not belong to the world at large, but to the company that paid for it, De Beers has attached the legend to its name on letterhead and advertising, clearly linking the two things that it sells—a mineral and an idea.

Charged since 1996 with keeping the idea bright is a bustling Irishwoman, Mary Walsh, a veteran of Yves St. Laurent and, before that, the Vendome Group, which owns such luxury brands as Dunhill and Mont Blanc. Walsh and her London staff work out of offices on Saffron Hill. Although connected to 17 Charterhouse Street, Saffron Hill keeps its own address, as if to demonstrate that the marketing of diamonds needs a breezier front than the selling of the rough around the corner. Walsh is short and speaks rapidly, brimming with confidence.

"Take a campaign like the millennium campaign," she said in late 1999, as the campaign was running. "We wanted women to start thinking 'diamond' and 'millennium' together. Men—we can get to them later. . . . It's women who are thinking [about the event] much earlier. So we aim our magazine advertising at the women early."

The first ads delivered what Walsh called "subtle" print messages to women, supported by pictures of women wearing diamonds. "It touches my hand gently," said the text, "as if to remind me that when I enter the next millennium I will not be alone." Or: "What else will be mine for a thousand years? What else will last as long as love itself?"

"See the diamond," said Walsh, drawing the words out in a stage voice and spreading her fingers wide as she described the purpose of the ads, "see the design. They're very girly, very female. You fuel her desire. If you don't talk to women you're never going to sell the damn thing. Now there's a bit of a dance going on here, because *she* has to approach *him*. This isn't just for the millennium, it's the same every year at Christmas. The nearer the event—millennium, Christmas, Valentine's Day—as you get closer to the event, the girl-to-girl is out. She's done. She's been pitched. Now we have to get to *him*." Walsh shakes her head. "That's so easy."

"Which millennium are you waiting for?" demanded one of the special cards mailed to men by De Beers in the last months of 1999. The message for men was more direct, the tone bantering. "Every thousand years or so it's nice to get her something really special for Christmas." The marketers even told him what to pay. In North America, an ad demanded: "Isn't she worth two months' salary?" This figure was adjusted according to a market study of what men in different parts of the world would pay. European men got off with one month's salary; the Japanese were asked for three.

Although De Beers spends more promoting diamonds than anyone else, the biggest boost the jewel gets is free. Hollywood loves diamonds. Sometimes the context is chilling, as when Liam Neeson, playing the lead in *Schindler's List*, heaped a pile of diamonds onto the desk of an SS officer in exchange for the lives of a trainload of Jews, or when Dustin Hoffman, in *Marathon Man*, flung diamonds onto a grate, handful after handful, and they fell through into a water-treatment tank as Laurence Olivier, playing a Nazi monster, scrabbled insanely to catch them. In these films the diamonds represented the overwhelming power of the jewel as a signifier of greed.

More often diamonds are a shorthand for another side of

desire—lust. In *To Catch a Thief*, Grace Kelly, playing an heiress, wears diamonds in only one scene, where she is trying to entice the man she suspects is a jewel thief preying among the women of the French Riviera. Kelly wants both to protect herself from the thief and to seduce him, a tension elucidated by the director, Alfred Hitchcock, when he lets a shadow fall across Kelly's face, erasing her individuality, so that the viewer's attention may rest completely on the diamonds and the female body, which form a single object of desire. A different, more blatant sexual statement is made in *Gentlemen Prefer Blondes*. Marilyn Monroe's song, "Diamonds Are a Girl's Best Friend," makes a series of simple connections. Men want sex; sexual attraction is a transitory asset for a woman; she had better exchange it for something with better liquidity.

In *Gentlemen Prefer Blondes* the jewelry used was paste, but stores like Cartier, Tiffany, and Van Cleef & Arpels have happily supplied real diamonds for movies. It is no accident that diamonds edged out other popular gems, such as pearls and emeralds, to become the world's favorite jewel. The great jewelers recognized Hollywood as a perfect showcase for their most costly wares, and they still do. When Whoopi Goldberg hosted the 1998 Academy Awards, she blazed away in $41 million worth of diamonds lent by Harry Winston, including a 107.18-carat ring worth $15 million. Geena Davis, the preshow host, sported a Winston ring worth $2.6 million and earrings that would sell for $45,000. "We made Geena a diamond bobby pin out of baguettes and rounds," a Winston spokesman said, "but the day before, she decided she didn't want the bobby pin. She wanted dangly earrings. So we broke the bobby pin in two and put in posts."

Gwyneth Paltrow, who won Best Actress that year for her role in *Shakespeare in Love*, wore a diamond necklace priced at $160,000, also on loan from Winston. It would be hard to imagine better advertising space than Gwyneth Paltrow's throat on Oscar night. After the ceremony the Harry Winston stores were inundated with calls. Gwyneth Paltrow's father bought the original for his daughter, but Winston's had twenty-five firm offers from buyers each ready to pay $175,000 for a copy.

◆

Diamonds draw on a mystique that goes back to ancient times. Alexander the Great, on his march into India, is said to have heard about a pit filled with diamonds. The pit was guarded by serpents whose gaze would kill a man. Alexander, eager for the diamonds, ordered that his men be given mirrors. When they approached the pit they held up the mirrors and turned the reptiles' gaze back on the snakes themselves, killing them. Alexander then ordered sheep to be slaughtered and their carcasses flung into the pit. The diamonds stuck in the fat. Vultures swooped down and devoured the diamond-studded flesh, and afterward, as they flew away, expelled a rain of diamonds into the hands of Alexander's men.

Diamonds were important enough to receive the close attention of the Indian general Chandragupta, who drove the Greeks from India in 322 B.C.E. and founded the first Indian empire. A treatise on statecraft prepared during his reign contained a chapter called "The Examination of Precious Articles to be Received in the Treasury." It listed the most important qualities of a diamond: a crystalline structure, brilliance, and size. By this time India had been producing diamonds for five hundred years.

The most celebrated diamond mine was at Kollur, in the old kingdom of Golconda, only a few miles west of the modern city of Hyderabad, at a place where the river Kristna cut a deep gorge through the rock. Many fabulous stones came out of the diggings there, and their fame spread through the world. When the French jewel merchant Jean-Baptiste Tavernier visited the site in the seventeenth century, he found sixty thousand men, women, and children laboring there. Two of the most storied diamonds in the world, the Hope Diamond and the Koh-i-Noor, came from the Kollur gorge. Today you can see the Koh-i-Noor in the Tower of

London. It is set in the center of the Maltese cross on the front of the crown made in 1937 for Queen Elizabeth, later the Queen Mother. Its brilliance led to the saying that whoever owned the Koh-i-Noor would rule the world.

The first reference to the Koh-i-Noor is in the *Baburnama*, a record of the life of Babur, the Mogul conqueror of India. In 1526 Babur marched against the sultan of Delhi, meeting his enemy at Panipat. Babur had an army of only twelve thousand against the sultan's force of one hundred thousand men and one hundred war elephants. A huge cloud of dust heralded the approach of the sultan's army. Babur, a descendant of Genghis Khan, opened the battle with artillery, then deployed his cavalry in swiftly wheeling formations that disrupted the antiquated battle order of the sultan. Babur routed the superior numbers, and in the course of the battle slew the sultan.

Among the princes who died with the sultan was the rajah of Gwalior. The rajah had a treasure of jewels, and before marching off to battle he had sent his family and his jewels to the Agra fort. But Babur found out about the rajah's treasure, and sent his son hastening to Agra to seize the stronghold. The rajah's family tried unsuccessfully to escape. When Babur arrived to receive the fort from his son, the rajah's family presented him with a *peshkash*, or placatory gift, and part of this treasure was the Koh-i-Noor. "It is so precious," waxed the *Baburnama*, "that a judge of diamonds valued it at half the daily expense of the whole world."

Writers often described the Koh-i-Noor that way: measured against the revenue of the entire world. It had no peer. The diamond passed from conqueror to conqueror, until in 1813 it came into the hands of Ranjit Singh, the Lion of the Punjab. Ranjit Singh was the first Sikh king, and the only one with real power. Three weak kings succeeded him. In 1843 the minor Dhulip Singh, the last of Ranjit's sons, took the throne. Six years later the British hoisted their flag in Lahore and annexed the Punjab. Like conquerors before them, they had an eye on the famous jewel, and the third clause of the Treaty of Lahore provided that "The gem called the Koh-i-Noor which was taken from Shah Shuja-ul-

Mulk by Maharajah Ranjit Singh shall be surrendered by the Maharajah of Lahore to the Queen of England."

The man responsible for guarding the Koh-i-Noor until it could be sent to London was Lord Dalhousie, the governor-general of India. The duty weighed on him, as he told a friend in a letter sent from India on May 16, 1850:

> The Koh-i-Noor sailed from Bombay in H.M.S. Medea on the 6th April. I could not tell you at the time, for strict secrecy was observed, but I brought it from Lahore myself. I undertook the charge of it in a funk, and never was so happy in all my life as when I got it into the Treasury at Bombay. It was sewn and double sewn into a belt secured round my waist, one end of the belt fastened to a chain round my neck. It never left me day or night, except when I went to Ghazee Khan when I left it with Captain Ramsay (who now has charge of it) locked in a treasure chest, and with strict instructions that he was to sit upon the chest till I came back. My stars! What a relief to get rid of it. It was detained at Bombay for two months for want of a ship, and I hope, please God, will now arrive safe in July. You had better say nothing about it, however, in your spheres, till you hear others announce it. I have reported it officially to the Court, and to her sacred Majesty by this mail.

In London, high expectations awaited the Koh-i-Noor. Some people murmured it would bring bad luck, and shortly after the diamond arrived at Buckingham Palace, a retired officer of the Hussars, apparently unhinged, struck Queen Victoria. The fame of the diamond grew. When the queen displayed it at the Great Exhibition in Hyde Park, crowds mobbed the structure that housed it and a near riot ensued. But some were disappointed in the jewel. The *Times* said the diamond failed to display the brilliance it was famous for. The *Illustrated London News* agreed, insisting that the diamond was "not cut in the best form for exhibiting its purity and luster," and ought to be recut. This was not surprising. Indian cutters in antiquity polished diamonds not for maximum brilliance but to preserve size.

Queen Victoria's consort, Prince Albert, took up the cause of polishing the Koh-i-Noor. The prince arranged for experts to examine the stone, and finally an Amsterdam firm was picked to cut it. A small steam engine was assembled at Garrard's, the crown jeweler, and the cutters arrived from Holland. On a Friday afternoon in the summer of 1852, the duke of Wellington, a keen admirer of the Koh-i-Noor, rode up to Garrard's on his charger. The Koh-i-Noor had been wrapped in lead, exposing only the first piece that was to be ground off. They fired up the steam engine that drove the grinding wheel. As the *Times* reported: "His Grace placed the gem upon the scaife, an horizontal wheel revolving with almost incalculable velocity, whereby the first angle was removed by friction, and the first facet of the new cutting was effected."

The Koh-i-Noor contained inclusions, and although experts had studied the diamond carefully, there were those who feared it would shatter into bits from the friction of the wheel. One of the Garrard brothers, asked what he would do if that happened, said: "Take my name-plate off the door and bolt." It took thirty-eight days to polish the diamond. The cutters reduced the Koh-i-Noor from 186 carats to 108.93 carats.

Prince Albert was dismayed at the loss of weight, and the news spread that the diamond's legendary ill luck had befallen those who had cut it. It was murmured that the stone had been fraudulently taken from the minor Dhulip Singh, that it had been his personal property and not the property of the state. As it happened, Dhulip Singh was in London at the time, posing for a portrait at Buckingham Palace. Queen Victoria sent to ask if the maharaja would like to see the newly cut stone. When this was put to him, Dhulip Singh said, "I should like to take it in my power, myself, to place it in her hand now that I am a man. I was only a child when I surrendered it to Her Majesty by the treaty, but now I am old enough to understand."

The next day the queen came into the room where the young

maharaja was posed on a dais, and handed him the Koh-i-Noor. He inspected it gravely, then handed it back. The report of this exchange enraged Lord Dalhousie, who maintained that the stone was no gift, but the Queen's by legal right. It is not surprising he was so insistent about ownership, given the story he repeated in a letter to a friend:

> When Ranjit Singh seized [the Koh-i-Noor] from Shah Shuja he was very anxious to ascertain its real value. He sent to merchants at Umritsir, but they said its value could not be estimated in money. He sent to the Begum Shah, Shuja's wife. Her answer was thus, "If a strong man should take five stones, and should cast them, one east, one west, one north, and one south, and the last straight up in the air, and if all the space between those points were filled with gold and gems, that would not equal the value of the Koh-i-Noor."

Every gem trader knows the value added by a story. Diamonds with provenance command a premium. In 1974, Harry Winston made a deal to buy $24.5 million in top goods from De Beers. He negotiated directly with Harry Oppenheimer. The transaction took the two only a few minutes. When they were through, Winston said, "How about a little something to sweeten the deal?" Without a word Oppenheimer reached into his pocket, fished out a 181-carat rough stone, and rolled it across the table. Winston smiled and picked it up. "Thanks," he said. As soon as he got back to New York he sent the rough to the wheel. It produced five polished diamonds. The largest was a 45.31-carat D flawless emerald cut, which Winston promptly christened the Deal Sweetener, and whoever bought it would have to pay for the story as well as the diamond.

Understanding the premium of names, some owners try to attach them. That is what happened to a 31-carat sapphire-blue diamond, cut into a heart and sold by Cartier in 1910 to an Argentine family. A rumor began that the diamond had belonged to Empress Eugénie of France. This could not have been true, because by the time the blue had been cut the empress had already been living in exile in England for almost forty years. Her circumstances were much reduced and her jewel-owning days lay far behind her. Nevertheless the rumor stuck, and when Harry Winston bought the blue diamond and sold it to Marjorie Merriweather Post, the cereal heiress, it carried the name Eugénie Blue. Mrs. Post gave the diamond to the Smithsonian Institution in Washington, where it still carries its deceptive name.

Not every owner of a large diamond succeeds at this game. Gary Schuler, an executive of the jewelry department at Sotheby's in New York, recalled a midwestern family who brought him a "named" stone. Schuler searched his reference books and could not find it. He refused to include it in his sale under the putative name. "They had all kinds of documentation, a whole stack of paper. You know, you get some local newspaper to do a story, and they just write down what you tell them. The family had all kinds of stuff like that—who owned it, who sold it, why it was called what it was called." Schuler would not disclose the name, shaking his head and smiling wryly. "Call it the Great Star of Ohio."

◆

When diamonds go to auction, more than money is sometimes at stake. The salesroom can become the site of a fierce and emotional contest, as on the evening of May 6, 1997, at Christie's old Park Avenue address in Manhattan, when Lot 171 came up to take its turn. The object was a brooch by Van Cleef & Arpels, replicating the flag of Argentina. Two outside bands of sapphires set off the middle strip of diamonds. The flag's central "Sun of May" was a fancy intense yellow diamond. The flag seemed to ripple bravely. It had, moreover, a glittering provenance—it had belonged to Eva Perón.

Bidding opened at the low estimate of $80,000 and got briskly under way. As competition heated up, all eyes in the room focused on a single bidder—a tall, slender, blond woman with a look of cool detachment. This was the Argentine actress Susana Gimenez. Normally, lots come up and are knocked down in a matter of a minute. But the bidding for Lot 171 soon passed beyond this into the territory auctioneers love, which is war. The merely covetous dropped out, leaving the field to two combatants—Gimenez and an anonymous telephone bidder, a collector in California, an individual whose passion for the object lay hidden behind the auction house's perfect sphinx of a young woman on the telephone desk, who took the bids and passed them on. Christie's had put a high estimate of $120,000 on the sparkling little flag, but it sailed past this without a pause, heading north.

Every time Gimenez bid, a murmur of approval went around the salesroom. Then every eye would swivel to the telephone desk, where the young woman sat stiffly at her place, eyes discreetly lowered, the phone at her ear. After a moment, she would raise her eyes and nod at the auctioneer. A wave of grumbling would sweep the room. Some people booed.

As the bidding passed $700,000 it settled into a grim contest. The price advanced in increments of $20,000. When it reached $780,000, Gimenez seemed to pause for a moment, as if to gather strength, before she tilted her chin and sent the price up to $800,000. The crowd howled and raised clenched fists. Then they turned to the telephones. The young woman's eyes were lowered barely a second before she looked up and nodded to the auctioneer. A sigh of dismay went through the room. Gimenez stayed gamely in as the price climbed inexorably up. Her last bid was $880,000. When the prompt reply came in—$900,000—she shook her head and the brooch was hammered down, and a dejected crowd watched wordlessly as the actress left the room.

A good provenance adds value to a jewel, but buyers and sellers generally insist on anonymity. Christie's would not say who outbid Gimenez for the Argentine flag, nor is it possible to extract from Simon Teakle, the head of Christie's jewelry department in New York, the names behind a remarkable treasure of jewels that came his way in England.

Simon Teakle with a close model of the Ahmadabad, a 78.86-carat top-color diamond acquired for Christie's by Teakle, and auctioned at Geneva for $4,324,554. (Christie's)

Teakle's adventure began on a spring day in 1989 when the telephone rang in the jewelry department at Christie's in London. No one knew the caller, but Teakle, a junior in the department, agreed to take the call. He is intensely genial, with bright red cheeks and sandy hair, a man who takes evident delight in his business and its stories.

"It was a lady," he said of the caller. "She had a pair of diamond-and-sapphire cuff links to sell. She wanted to know how the auction process worked. I gave her the nuts and bolts—send in the goods ten weeks in advance, commission of ten percent, et cetera. She said she had no plans to travel to London, so I suggested she send them in by registered mail [a common insanity of the trade]. She was worried they'd get lost. I asked where she lived, and as luck would have it I was staying that very weekend with friends about ten miles away."

On Sunday after breakfast, Teakle got into his primrose yellow BMW and went spinning off through the lanes of Kent. The directions took him to a modest eighteenth-century house. He sat down in a sunny room furnished with large, chintz-covered sofas, and helped the lady and her husband finish off a pot of tea. An hour later he drove away with the cuff links, sold them for £3,500, and heard no more for six months.

"Then I got another call. 'We have a blue diamond we'd like to sell.' I knew right away what they'd been up to with the cuff links—they were testing the auction process. A blue diamond, now that's a very different animal. I knew she wouldn't put it in the mail so I drove out right away. When I got there I saw one of the most beautiful blue diamonds I'd ever seen, set in a pendant. I was still quite inexperienced, and when she asked me what it was worth I took a very deep breath, reached down for as high a figure as I thought it could get, and said five hundred thousand pounds."

Lot 551 went into the catalog for the autumn sale as "The property of a lady." Christie's described it as a "magnificent belle époque blue, orange-yellow, and colorless diamond pendant." The stones were classed as fancy dark blue, fancy intense orange-yellow, and E-color whites, VS2 (very slightly included). It sold for £1.5 million. "They were not an extravagant couple," Teakle says, "but she had a mania for caviar. The day after the sale I got back to my office and there was a bucket of the stuff on my desk."

One year later, she called again. "We're very happy with you, Simon," she said. "We are happy with Christie's. My husband and I are getting on, and we have some things that perhaps we ought to sell."

"Well, I *tore* up there," said Teakle. "They had no heirs, only charities, and they wanted to leave cash." When he got there he found the Waterfall, a necklace of D-flawless white diamonds, a piece that by itself fetched just under £1.5 million. The whole collection brought more than £10 million. Teakle would not identify the owner, except to say that she was the illegitimate daughter of one of the great diamond families of the late nineteenth century.

That same year fortune smiled on Teakle again. A family in Scotland owned a famous diamond. Christie's had been angling for the commission for years. When the call came through, Teakle flew to Edinburgh and drove to a provincial bank some fifty miles away. A strongbox was produced and opened, and in it lay a 13.5-carat rose-pink diamond that Babur had worn in his turban. Known everywhere as the Agra diamond, the pink jewel had been part of the treasure that included the Koh-i-Noor, which Babur took from the widow of his vanquished enemy. The Agra had passed to subsequent Moguls, until it disappeared, not surfacing again until 1844, when the duke of Brunswick bought it in Paris. In 1891 the London dealer Edwin Streeter bought the Agra, which had by then been cut to its present size. Streeter's business successor sold the pink diamond to a dealer, who resold it to Louis Winans, the son of an American entrepreneur who had built Russia's first commercial railway. Winans's heirs kept it for sixty years before sending it to auction. "A magnificent unmounted cushion-shaped fancy light pink diamond" was how Christie's put it in the catalog. A horde of film stars, tycoons, and royalty came to gaze at it. Christie's sold it in London at the first evening jewelry sale to be held at its King Street rooms. It brought £4.5 million.

◆

The auction of famous jewels adds to the perception of value, for if the rich will pay large sums for them, they must be worth it. But nothing brightens the charisma of diamonds so much as people stealing them. Thieves are great promoters of diamonds, whether they clean out a store at gunpoint or steal the necklace from a countess's throat or seize some historical moment.

On the night of June 20, 1791, Louis XVI, the French king, fled through the night with Marie Antoinette in a desperate bid to escape the revolution. Troops caught the king's carriage at Varennes, and the king and queen were returned to Paris under guard and imprisoned in the palace of the Tuileries. Civil order rapidly declined. The government decided to bring the kingly accoutrements under the same watchful eye as they had the king himself, and ordered the crown jewels transferred from Versailles to Paris. The treasure moved out of the great palace in a thicket of muskets and rattled northeastward to the capital. Some of the most celebrated diamonds in the world went down the road that day, but none more famous than a steely blue heart-shaped diamond of 67.13 carats, the French Blue.

The diamond had been in France for more than a hundred years. Tavernier had sold it to Louis XIV in 1669. At that time it weighed 110 carats. Tavernier described it as *un beau violet*—which meant a blue of great intensity. The Sun King covered himself in diamonds. Indeed, he shone like a kind of sun when he appeared among his courtiers in garments sewed with diamonds from head to foot. Even his shoe buckles blazed with diamonds. Diamonds sparkled on his breeches. One coat had sprays of diamonds stitched to it, and 123 diamond buttons. The buttonholes too were sewn with diamonds, and beneath the coat, a jewel-spattered waistcoat glittered with the gems. A velvet hat had seven diamonds in it. With such a love of radiance, it was inevitable that Louis would decide that the large blue, cut in the Indian style, must be improved.

In 1672 Louis sent it to his jeweler, Pitau, who polished away some 50 carats and produced a beautiful heart-shaped diamond that dazzled the court. The king wore it at his throat on great occasions. They named it the Blue Diamond of the Crown. Its fame spread through Europe, where connoisseurs called it simply the French Blue. The Sun King's successors prized the diamond too. Louis XV had it set in the Order of the Golden Fleece. When he married, he took it out and gave it to his queen, Maria Leszczynska, who wore it in her hat. Such was the reputation that preceded the diamond as it rumbled along the rutted road from Versailles to Paris.

The security of the royal treasure rested in the hands of an officer of the court, Thierry de Ville-d'Avray. He agonized over the safety of the jewels. Public order was in a bad state. The government ordered him to put the treasure into the Garde-Meuble, a state warehouse filled with works of art and furniture from the royal palaces. But the Garde-Meuble was open to the public every

Monday, which worried the old courtier. A thief could study at leisure the layout of the building, view the jewels, and even inspect the caskets where they would be locked away. One thief did just that. Paul Miette, a thirty-five-year-old felon, previously banned from Paris, had returned to the city in the growing confusion. Records show that he visited the Garde-Meuble on open days. He was arrested in March of 1792 for a theft in the city, and imprisoned.

The security of the Garde-Meuble rapidly decayed. Sentries rotated according to a vague, unspecified system of shifts. On June 20, tormented by worry, de Ville-d'Avray packed all the unmounted gemstones into boxes and carried them to his private apartment. He stuffed them into an alcove and concealed them behind a heap of parcels. When the revolutionary leaders found out, they ordered the jewels returned to their previous place, and that is where they were on August 10 when a mob stormed the Tuileries and looted the palace.

Things quickly fell apart. On September 2 the populace rioted again. They broke into the prisons and massacred the noblemen and clergy interned there. They set the common prisoners free, including Paul Miette. A slaughter of the upper class ensued, and de Ville-d'Avray perished. His replacement found the Garde-Meuble in a pitiful state. The ground-floor windows were not even barred. He reported to his superiors that often there was no guard at all. The greatest treasure in Europe lay waiting to be plucked.

The sack of the French crown jewels began on the night of September 11. Two bands of thieves met at eleven o'clock, the first led by Miette, the second by a scoundrel from Rouen named Cadet Guillot. While some of the thieves masqueraded as guards in what is today the Place de la Concorde, the more agile scaled the cornerstones of the Garde-Meuble and gained a balcony on the second floor. They found the shutters unbarred. Breaking them open, they cut an armhole in a windowpane with a glazier's diamond. One man reached inside and flipped the window latch, and they were in.

That night they took the Sun King's sword and watch and the jeweled buckles of his shoes. They broke open a chest and found the colored parure and the white parure—assemblages of jewels mounted together. Miette took the white parure, which contained the Second Mazarin, a large table-cut diamond. The colored parure, snatched by Guillot, blazed with two of the most cherished diamonds of the court—the Côte-de-Bretagne, cut into the shape of a dragon, and the French Blue.

The next night the thieves brought supper. When they had finished stuffing their sacks with treasure they sat down to a meal of bread and sausages and wine. A mood of recklessness spread through the enterprise. By September 16 the gang had swollen to more than fifty. Some of them donned uniforms of the national guard and posted themselves in the square to protect the plunderers. As before, the thieves entered through the second-story balcony. By midnight they were done, and clambered back down to the street. A dispute broke out when some of the party demanded an immediate division of the spoils. They made such a racket that a patrol of guards in the rue St.-Honoré heard them, and came to investigate. The thieves' accomplices sounded the alarm and the robbers fled.

Police quickly broke the gang, recovered many diamonds, and guillotined some of the culprits. But the French Blue disappeared. It took scholars two hundred years to trace what happened to it. Its path was strewn with false leads and suppositions. A portrait of the Spanish royal family, painted by Goya in 1798, showed Queen Maria Luisa wearing a large, dark blue jewel that could have been the French Blue, recut by the thieves to disguise its provenance. This theory has been discounted. It seems that the diamond stayed with Cadet Guillot. He fled from Paris to Nantes, and from there to le Havre, and finally to London.

At the end of the eighteenth century, London diamond dealers had three sources for their gems—India, Brazil, and the French Revolution. Yet the French Blue would be hard to sell. It was then the most famous jewel in Europe. A buyer would have had to fear that the French government would one day claim it back. When Guillot sold it, then, its fate was sealed, although it must have broken the heart of whoever put it to the wheel. For that is what happened. The blue heart was polished away forever. Gem experts now agree that a certain blue diamond that appeared in 1812 in the hands of Daniel Eliason, a London diamantaire, was the recut French Blue. A lapidary who made a drawing of the stone described it:

> The above drawing is the exact size and shape of a very curious superfine deep blue diamond. Brilliant cut, and equal to a fine deep blue sapphire. It is beauty full and all perfection without specks or flaws, and the colour even and perfect all over the diamond. I traced it round the diamond with a pencil by leave of Mr. Daniel Eliason and it is as finely cut as I have ever seen a diamond.

Probably the writer cut it himself, but could not take credit for the act for reasons of law, and probably also from shame, for no matter how well he cut it he obliterated one of the fairest jewels in the world. A legend of bad luck has followed it ever since. A Russian prince is said to have given the diamond to an actress of the Folies Bergères, then shot her across the footlights in a jealous rage. The jewel hung at her breast as she died. Or that is the story. What we do know is that the heart-shaped blue diamond that came out of the Garde-Meuble on a September night in 1792 weighing more than 67 carats was bought in 1830 in London by Henry Philip Hope for $90,000; it weighed 44.5 carats. The Hopes had the jewel for thirty years before gem experts conclusively established that the diamond was the lost French Blue. By then the family name of the Hopes had become attached to the stone, which was always thereafter called the Hope Diamond.

In 1901 Lord Henry Francis Hope, his fortunes in decline, sold the diamond to a Hatton Garden dealer. In 1928 Hope succeeded his father as duke of Newcastle, and such was his chagrin at having sold the great diamond that he would never permit anyone to speak of it in his presence. When he died in 1941, his obituary carried a heading that would have irked him: "The Duke of Newcastle. Former Owner of the Hope Blue Diamond."

After the Hopes, the diamond belonged to Abdul Hamid II, the Ottoman sultan, who was forced to sell it when a revolution drove him from his throne. Pierre Cartier bought the blue in Paris, and in 1910 showed it to Mrs. Evalyn Walsh McLean, an American mining heiress and the wife of a newspaper tycoon. Mrs. McLean did not like the diamond, but Cartier would not give up. He set it in a necklace blazing with white diamonds, took it to New York, and sold it to Mrs. McLean for $180,000. Years later when Mrs. McLean died, on a Saturday, her executors were so nervous about the safety of the famous jewel on a weekend with the banks closed, that they appealed to J. Edgar Hoover, the director of the FBI. Hoover allowed them to place the Hope in a safe at the FBI headquarters building.

Harry Winston bought the Hope diamond in 1947. He paid $179,920, then insured it for $1 million. In 1958 he presented it to the Smithsonian in Washington—the most beautiful blue diamond in the world, surrounded by sixteen white diamonds and fastened to a necklace of forty-five white diamonds. In more than forty years at the museum, the Hope diamond has drawn more visitors than any other display or object.

8
Bad Goods

Malfeasance rustles in the background of the diamond world like a snake in dry grass. Someone is always stealing diamonds. A circus of swindling and larceny enlivens the trade, and even those who live from diamonds sometimes fall victim to the miscreants.

In January 2000, a Chicago jewelry salesman pulled his Lincoln car over to the side of the road when a loud noise came from the engine. With the engine still running, he got out to see what was wrong. As he stood in the road, two men appeared and leaped into the Lincoln, slamming the door in the salesman's face and roaring away. In cases under the seat were $1 million worth of polished diamonds. Police still do not know exactly how the robbers engineered the sudden noise from the car.

Sometimes it is the jewelers themselves who steal. The same

month as the Chicago heist, two American jewelers went on trial charged with stealing $80 million from their customers, either by selling them fakes or by replacing with fakes the high-end gems their customers sent in for cleaning. But all this pales beside the wholesale plunder of the mines themselves, nowhere carried out with such panache as on the Diamond Coast.

In the northwestern corner of South Africa lies a slab of desert called Namaqualand. The Atlantic Ocean crashes on its coast and the beach is often lost in fog. The Diamond Coast begins in Namaqualand and runs north into Namibia. Diamonds arrived on the coast 15 million years ago, tumbling along rivers and into the sea. As in all alluvial deposits, river currents carried off the smallest diamonds, leaving caches of larger gems that were heavy enough to settle into place. That is why alluvial deposits produce higher per-carat values than would be found in a typical mine—the rich trove of alluvials is not adulterated by the lower values of tiny stones.

When the ancient sea receded, the diamonds that had flowed into it remained behind in beach deposits. Others lie farther out, embedded in gravels on the ocean floor. Beach and sea, a great swath of diamonds trails up the desolate coast.

At Kleinsee, where the Buffels River meets the ocean, De Beers takes 700,000 carats a year from the beach. North of Kleinsee lie the great concessions of Alexkor, South Africa's state-owned mine. Alexkor owns the beach from Kleinsee to the Orange River and maintains a flotilla of inshore diamond boats. And across the Orange River, taking up the entire southwest corner of Namibia, is a sprawling tract of sea and beach and thirty-two thousand square kilometers of surrounding desert, the whole of it identified on maps as Diamond Area 1. The maps advise that travel there is prohibited. Diamond Area 1 is the property of Namdeb Diamond Corporation, a consortium of De Beers and the Namibian government.

The borders of Diamond Area 1 are exactly those of the old *Sperrgebeit,* the "forbidden territory" proclaimed by the region's former German masters when diamonds were discovered in the early years of the last century. Diamond Area 1 was once the richest diamond ground on earth. It retains a special place in the hearts of those who live in Namaqualand, because it helps to water their arid province with the only rain that counts—stolen diamonds.

In Namaqualand, stealing diamonds is the proper work of man. Frikkie Mostert, an old diamond hand, seeking to explain this, drove his pickup down into the delta of the Orange River. Tall grasses grew in the brackish water and a flock of lesser flamingos stilted through the shallows, their pink plumage shimmering in the heat. The deep, rhythmic boom of waves against the beach reverberated through the delta. Mostert parked the pickup and got out.

Half a mile away, in Namibia, a truck drove slowly along inside the wire fence that marked the border of Diamond Area 1, and therefore of Namibia. Beyond the fence, in the distance, rose up the first of a line of dun hills that range along the coast. The hills are mine tailings, great ridges of gravel stripped from atop the diamonds and piled by the sea. In the other direction, south of the river, lay the neat establishment of Alexkor's company town, Alexander Bay, and beyond it the long, hot, bleak stretches of the mine property. There was really nothing but the ocean and the desert, the one frigid and veiled in fog, the other parched by the relentless sun. Mostert gestured around with his hands, and he had a worn smile on his face, and he said, "God put the diamonds here and He put nothing else. Ach, man, people here think the diamonds belong to them. Stealing diamonds, it's not stealing."

Beach mines are easy to rob. In other types of mines, such as open pits, miners rarely come face-to-face with diamonds. Earthmoving equipment scoops out the ore and feeds it to the recovery plant. But on the beach, diamonds lie on the old seabed. Miners strip away overlying gravels, sweep the bedrock with whisks, and search every crevice for gems. Namdeb has introduced vacuum machines to make it more difficult for miners to get their fingers next to diamonds. The men get them anyway. As the axiom goes, if they can see and touch it, they will try to steal it. A foreman's

Vacuuming rough at Namdeb on the Diamond Coast.
(De Beers)

eyes cannot be everywhere at once, so a miner flicks a stone up his sleeve or stamps his boot down hard enough to make a diamond stick in the sole. The gem can be picked out later and pressed into the tire of a vehicle leaving the beach. An accomplice in the maintenance yard will retrieve the stone. But the hallmark caper of the Diamond Coast was the pigeon scheme.

For a time in the late 1990s, it was common for miners at Namdeb to keep homing pigeons in their hostels at the mine. Managers did not object. Here were numbers of young males, separated from their families for long periods: if they could busy themselves with some harmless birds, then fine. The problem with allowing this pastime, as management would discover, lay in the fact that the homing pigeons' homes were not the mine but places outside the mine and, unlike the miners, the birds could return to them without passing through mine security on the way.

To begin, the miners trussed the birds tightly and smuggled them out to the beach in lunchboxes. Then, fitted with tiny harnesses, the pigeons were loaded with rough and set free. They fluttered up into the air, circled the sprawling digs a few times, and flew off to their homes. There they were tenderly received, and their little harnesses emptied of diamonds. As soon as the time was right, back they went to the mine for another run. A good deal of rough made its way out of Namdeb in this fashion, and in Namaqualand the phrase "the birds are flying" meant that a steady supply of contraband was coming out of Diamond Area 1. The system might be flourishing still had not a miner gotten too greedy, and loaded his pigeon with so much rough the bird could not take off. As it dragged itself along the ground, beating its wings in the dirt, the pigeon caught the eye of a security officer. The birds were quickly banned, and pigeons are now shot on sight by mine guards all along the Diamond Coast.

Soon the birds were replaced by something else: arrows. Miners smuggled in the pieces of a crossbow. They hollowed out the shafts of the arrows, filled them with diamonds smuggled off the beach, and at night fired them over the fence to be picked up by cohorts. De Beers does not know how long this shuttle sent its freight of rough over the wire, but the scheme ended when an arrow hit a security jeep that was making a patrol along the fence.

Diamonds get dropped into gas tanks, wedged behind sweatbands, tapped into ears. At one De Beers mine, security officers caught a thief only because he'd inserted so much rough into his rectum he was waddling. Diamonds fluoresce under X ray, and X ray scanners, where they are allowed by law, are used to examine miners, although a frequently corrupt security force can and does interfere with the machines.

At Namdeb, the biggest problem comes from an historical anomaly: the existence of miners' hostels inside the secure perimeter of the mine. A large population with lots of time to think about stealing diamonds permanently occupies the most sensitive area of a famously rich mine. Criminal syndicates, active at mines throughout the region, conduct a brisk diamond bourse inside Namdeb's hostels. De Beers has admitted this. Said Vice Admiral Sir Alan Grose, a former De Beers security chief: "The diamond industry has tended to attract all kinds of unsavory characters, not just to steal, but to intimidate people, to introduce a general criminal atmosphere. Once that gets in place, it's very hard to eliminate."

Nor could Namdeb raid the hostels. The syndicates contained veterans of the same organization—South West Africa People's Organization (SWAPO)—that constituted Namibia's ruling party. Put another way, the government's former comrades-in-arms were looting the state's main asset, and for reasons of political delicacy, and perhaps of corruption, the predation could not be stopped. Unable simply to burst into the hostels, Namdeb's beleaguered security tried a different tack: a raid on the buses that carried miners to the beach and back.

"The thinking was," said a former De Beers mining executive, "if we can't do anything about the hostels, let's stop the buses on the way back. That way we wouldn't be infringing on sensitivity about invading hostels and the miners would still be on shift."

Without warning, security stopped the buses. They found nothing, and had to let the buses go. As soon as the vehicles pulled away, the raiders noticed a litter of tiny parcels on the road a hundred yards back. They were full of rough.

Once in the hostel, stolen diamonds had another hurdle to cross: the fence. Namdeb subjected miners to X rays. Because of the health danger posed by accumulated X ray radiation, such X rays were random. Sometimes a miner would gamble on the randomness and carry diamonds through the checkpoint, knowing that if he was caught and sent to prison, the syndicates would care for his family. De Beers then developed a low-dose X ray called Scannex, which allowed the security staff to use the X ray more frequently. In September, 1999, they caught an employee trying to pass the checkpoint with 51 diamonds worth $2.5 million in his stomach.

◆

Namdeb's security concerns extend far past the fence, to the beach, the shore itself, and out to the pitching seas of the Atlantic. De Beers has been mining the seabed for almost forty years, since it bought a controlling stake in the pioneer ocean-diamond-mining venture of Sam Collins, an irrepressible Texan who had heard about ocean diamonds from a naval officer and adventurer in Cape Town. This man in turn had been stimulated by the tales of a South African farmer who had put a wooden fishing vessel into the turbulent inshore waters of the Namibian coast and sucked up diamond-bearing gravel with a pump. Single diamonds had been found for years in shallow waters from Angola to South Africa. That these occasional gems might point to substantial deposits was suggested by the presence of fossilized oyster shells in the onshore diamond gravels. If the beach deposits had once been covered by the sea, thought Collins, why would there not be more deposits, equally rich, on the present seabed?

Collins bought the farmer's marine concession, formed a company, and in 1961 the first of a ragged succession of ships and mining barges sailed north from Cape Town. For the next few years, Collins's vessels struggled in the heaving waters. Banks of dense fog often rolled in and blanketed the operation. The infamous storms that pound the Namibian coast tore the barges from their moorings and drove them onto the rocky shore. But they found diamonds. In the first five days they sucked up 2,100 stones with a total weight of 1,018 carats. The rough was of excellent quality, and at today's prices the parcel would fetch around $300,000. Another haul yielded 1,100 carats in six days. Collins bought a better ship. An aircraft replaced the boat that had ferried crews back and forth along the coast; now a fresh crew could be flown in from Cape Town in little more than an hour. With his new resources in place, Collins's production leaped. One six-day period yielded 4,000 carats. But the treacherous conditions of the Diamond Coast sometimes stranded the Texan's ships in port. Heavy financial burdens pressed the company: its payroll had swollen to five hundred, and the costs of keeping its fleet at sea exceeded the improved but often interrupted income. Finally, in 1965, De Beers moved in and bought control.

De Beers threw itself into the project with a fury. The pride of Sam Collins's fleet, the diamond ship *Diamantkus*, steamed back to Cape Town. De Beers removed the ship's twin 4,400-horsepower engines and installed them as power generators on a massive barge, the *Pomona*, that was rising in the shipyards. The *Pomona* was 285 feet long and 60 feet wide. Above her steel decks climbed a towering contraption of girders and pipes and chutes and winches. A five-story structure at the forward end had bunks for a crew of 120, and one whole deck for recreation. A helicopter landing pad was laid out on the roof. The vessel had an ungainly, top-heavy appearance, as if she must keel over and sink at the first slap of a wave. The *Pomona* was completed in June of 1967, and with hopes of 50,000 carats a month dancing in her captain's head, the barge was towed out of harbor and up the coast. But the hazards of the inshore waters defeated De Beers, too, and in 1971 the company scrapped the inshore operation and set its sights on deep water.

Today the crew change for the ocean fleet arrives at Alexander Bay by plane from Cape Town. A huge yellow passenger helicopter waits on the hot tarmac. When all are aboard, the helicopter lifts away from the baking surface of Namaqualand, crosses the delta of the Orange River, and clatters out over the ocean. One fall day in 1999 the sea was a cloudy green trimmed with whitecaps by the breeze. Sun sparkled on the surface of the heaving ocean. The fleet was working the seabed fifteen miles out, and the helicopter I was in butted along through the buffeting wind for ten minutes before the first of the red ships appeared. Soon the whole squadron was in view, five scarlet-hulled vessels with white topsides, widely dispersed on the tossing sea. The helicopter began its approach to the *Debmar Pacific*. The landing pad protruded from the stern of the ship, extending out over the water and supported by struts. It pitched and rocked with the motion of the ship. The helicopter pilots maneuvered carefully above the painted target on the pad, waiting for the ship to rise up on a swell, then settling smartly down. Crewmen rushed out and secured the aircraft to the deck.

That day the ship was mining in four hundred feet of water. A drill rig towered above the superstructure. The drill shaft

A De Beers mining ship off Namibia. (De Beers)

plunged straight down through a gaping square hole cut into the center of the ship. As the ship rose and fell, green water surged up into the cavity and lapped at the decks. In the center the great drill turned ceaselessly. On the ocean bottom a massive steel drill bit, twenty-two feet in diameter, smashed the hardened gravels. Pumps sucked up the sludge and broken rock from the seabed into the ship's recovery plant. A steady stream of mud and rejected boulders cascaded over the side and into the green sea, and a milky brown stain trailed away on the current.

The shipboard recovery plant was a miniature of the bigger plants ashore. Diamonds are recovered from ore by a simple process that begins with the dumping of boulders too large to be crushed. Next, lighter materials are separated out and discarded. What remains is a concentrate of heavy minerals, including diamonds. This concentrate proceeds to the final recovery stage, where the minerals slide down narrow chutes onto a moving conveyor belt. As the stream of minerals reaches the end of the belt and begins to tumble off, it is bombarded with X rays. A photomultiplier is poised to detect the fluorescence that diamonds emit under X ray. When the device registers a burst of fluorescence from a diamond, it triggers a jet of air that knocks the diamond from the falling stream of gravel into a bin. The system is remarkably effective, and among the computer screens of the *Debmar Pacific*'s recovery control room, the ever-present background clanging of the plant was relieved by the intermittent popping sound that signals another diamond being blasted out of the concentrate and into the coffers of De Beers.

Large crushers and cyclones and vibrating steel grates form part of the apparatus of recovery. Altogether, a large mass of violent machinery bangs away at the ore, subduing it into concentrate. As one would expect in such a din, breakdowns occur. Any such event presents an opportunity to steal goods, for the simple reason that it puts human beings in close proximity to rough. Screens that normally keep people away from the recovery stream must be removed. Because breakdowns permit access to diamonds, employees engineer accidents. This in turn may require a repair technician to fly out to the ship, increasing traffic on and off the vessel. Increased movement of personnel exacerbates an already problematic situation, namely, the fact that it is easier to remove stolen goods from a ship than from a typical mine. As Sir Alan Grose has pointed out, the normal activity of servicing a ship at sea generates more opportunities for criminal collusion.

In 1999 the red ships added some 485,000 carats to Namdeb's production. The inshore fleet—private contractors with forty-foot boats—kicked in another 135,000 carats. The beach yielded 786,000 carats. Altogether, Namdeb produced almost 1.5 million carats worth more than $400 million. This total would have been much higher were it not for some $180 million worth of top rough that was robbed from Namdeb that year—30 percent of the mine's production. Probably another $20 million worth of stolen diamonds came out of Alexkor, the state-owned diamond mine south of Namdeb, on the South African side of the Orange River. Alexkor was a laughingstock along the Diamond Coast, and nowhere was the laugh enjoyed so much as in Port Nolloth.

◆

One of the most enduring staging posts for stolen rough from the Diamond Coast is Port Nolloth, a dreary coastal hamlet in northern Namaqualand, some ninety kilometers from the Namibian border. A cold wind blows from the ocean into the dusty streets. The town dock is an oily, L-shaped structure with a crane on rusty rails. A little fleet of roundish, fiberglass diamond boats, called "tupperwares," bobs in the shelter of the reef. At the south end of town is a small recovery plant where Trans Hex processes diamond-bearing gravels from the inshore waters.

The Diamond Coast is a stormy coast, and often the sea is so violent that the tupperwares remain in port. In the early morning, divers bicycle out of the fog and arrive at the dock. They stand about and stare at the heaving ocean and mostly turn around and pedal back home. On days when the wind subsides, the fleet goes rolling out to sea, one boat behind the other, their suction hoses trailing behind them. When a boat reaches its concession, divers go over the side into the frigid waters of the Benguela Current. They descend into the dangerous, murky surge.

On the seabed, they wrestle steel nozzles into the gravel. Divers are often injured, their ribs cracked when boulders, hidden in a haze of silt and undermined by the gravel-sucking hoses, suddenly roll onto the unsuspecting men. Violent seas slam them against the hull as they climb from the water. Many suffer chronic joint problems from diving too deep for too long, and not decompressing when they leave the water. Burst eardrums are common. Often the tupperwares are stormbound for all but a few days a month, and in the grimy Scotia Inn the divers sit and drink the nights away. For men with such a life, stealing a diamond here and there would be a natural temptation. But the small-time thief is not the source of Port Nolloth's notoriety.

On the Diamond Coast it is an article of faith that rivers of illicit rough flow through the little town. Despite its lack of credible signs of commercial activity, Port Nolloth possesses a flourishing population of BMWs and Mercedes-Benzes. At the north end of town is a cluster of concrete villas, painted ocher or white, which stand by themselves and shoulder aside the encroaching dunes. They suggest an anomalous affluence, there between the empty desert and the sea. When the Gold and Diamond Branch of the South African Police decided to raid Port Nolloth in 1995, the raid fell hardest on one particular group of the town's residents—the Portuguese.

On the appointed day, the police gathered their force at the De Beers airstrip at Kleinsee, south of Port Nolloth. At a signal, vehicles loaded with heavily armed officers roared out of the compound and sped north. A helicopter clattered into the air and headed for the main target, the Portuguese country club on the highway south of Port Nolloth. Walls topped with razor wire surrounded the facility. The helicopter racketed over the wall and hovered. Combat officers slid down ropes into the green oasis of the grounds. They charged the clubhouse. Later, recounting the assault, a diamond detective with more than twenty-five years on the force shook his head and recalled finding diamond scales and loupes. "Only there weren't any diamonds. Later we raided a house, and found two hundred fifty thousand dollars cash."

"Yes," his young superior added with a sigh, "but the possession of money is not illegal in South Africa."

The Diamond Coast is a perfect sieve of leaking rough, but theft afflicts all mines. Botswana is often pointed to as an example of how greatly diamonds can benefit a country. A cursory glance seems to support this affirmation. Gaborone, the capital, is a fresh little city of about 150,000. Schoolchildren in clean uniforms wander arm in arm along the streets when school is out. The town is planted with trees and flowers. Boulevarded avenues carry traffic smoothly around the perimeter. The hotels are full of prosperous foreigners meeting prosperous Botswanans, and a

Rolls-Royce purring into the parking lot does not stir much interest. But the country has a population of 1.5 million, and they are not all packed into the hotels eating shellfish from the cape. Many people in the countryside have no work.

The malign influences of the outside world have not ignored Botswana. Criminal syndicates in neighboring South Africa have thoroughly penetrated Botswana's diamond mines. To prevent losses that may have been running at $70 million a year, the Jwaneng mine opened in 2000 a new ultrasecure recovery plant. De Beers engineers nicknamed it Aquarium, from the acronyms FISH (Fully Integrated Sort-House) and CARP (Completely Automated Recovery Plant). The aim of the improvements was to better keep employees separated from the stream of rough, it being axiomatic in Botswana, as everywhere else, that anyone who can actually touch a diamond will try to steal it.

The culture of brigandage that afflicts large parts of southern Africa might seem to explain the robbery from mines. But mines are robbed everywhere, as executives of the Argyle diamond mine in Australia discovered. Argyle lies in remote country at the edge of the Great Sandy Desert, some twelve hundred miles northwest of Western Australia's capital, Perth. The mine's production consists mostly of cheap brown goods. The exception is a sizable quantity of pinks, which Argyle markets separately. "Basically some of the diamonds popped up in Europe," said an Argyle spokesman. "That's how we mostly became alerted to the fact [of the theft]. We are really the only major world supplier of pink diamonds. They were cropping up overseas. They were our sort of goods, and yet we had not sold them." The Australian press fell upon the story with gusto. As the tale unfolded, a perfect cast of villains emerged—an embittered lover, her cuckolded husband, and the dashing, down-on-his-luck entrepreneur who cooked up the scheme.

Lindsay Roddan, a businessman in Perth, the capital of Western Australia, was a former schoolteacher who had built a reputation as a high-flying entrepreneur with a fortune in real estate and horse breeding. But his ventures failed, and by 1989 he faced ruin. At about this time he began an affair with Lynette Crimmins, a thirty-eight-year-old married woman. With Lynette Crimmins as his accomplice, Roddan set out to convince the Argyle security chief, Barry Crimmins, Lynette's husband, to steal pinks.

Roddan told Crimmins that all the man had to do was get the diamonds, and that he, Roddan, could dispose of them without a trace. Crimmins's "moment of madness," as he called it, arrived when four quart-sized containers of pink rough landed on his desk. Diamond mines wash their production in acid to remove coatings that dull the appearance of the goods. In this case, Argyle's sorters in Perth had sent back containers of high-end goods, including pinks, asking that they be washed again. When the rough went out to the cleaning plant in the nearby town of Kununurra, the couriers somehow forgot one of the containers. When Crimmins discovered the mistake, he checked the couriers' manifest. There was no mention of the forgotten container. He stole it, and the robbing of Argyle got under way.

Roddan paid Barry Crimmins in cash, sometimes as much as U.S. $10,000 at once. Some of the pinks went straight out of the country. Others were polished in Perth, then smuggled out by a network of airline employees, who hid the diamonds in jars of face cream and carried them overseas in their personal luggage.

As the stones began to appear in Europe, Argyle's executives became convinced the mine was being robbed, but did not know how. They tightened their own security, and reported the theft to police in Perth. The Western Australia State Bureau of Criminal Intelligence conducted an inquiry. They concluded that a ring was stealing pink rough and selling it in Geneva and Antwerp. The case went from the bureau to the state police.

After a short investigation, a senior officer reportedly said, "Police cannot find any evidence that Argyle diamonds have been smuggled out of the country." It was a bare reply, especially in the face of a report from Argyle's auditors, Coopers & Lybrand, that diamonds did indeed appear to be missing, and that an important diamond register had been improperly maintained. Besides, Argyle's executives knew their own goods, and those goods were turning up in Europe, and Argyle had not sold them. They forced the state police to reopen the case.

After a second inquiry, the police again told Argyle there was not enough evidence to bring a prosecution. Furious, Argyle hired its own investigators and launched a civil action to find out who owned a parcel of diamonds seized by police, a parcel Argyle suspected had been stolen from the mine. A third inquiry opened. Argyle, already paying its own investigators, undertook some of the costs of the police investigation, too. "We felt so strongly that we funded further investigations," Argyle stated, "and as a result of those investigations police internal affairs became involved."

Finally Lynette Crimmins broke. She told a story of a love affair that had degenerated into hatred. When she appeared in court to plead guilty to conspiracy charges, her lawyer said Lynette Crimmins had been forced into prostitution by Roddan when the money from the diamond thefts dried up. Barry Crimmins, who pleaded guilty alongside his estranged wife, said he had been driven into crime by her greed. Lindsay Roddan's lawyer dismissed Lynette Crimmins's story as the tale of a woman hungry for revenge on a lover who had passed her over. In February 1994, a judge of the Perth District Court sent Barry Crimmins to jail for four years. Lynette Crimmins, for having cooperated with police, got three years' parole. Two years later Lindsay Roddan was sentenced to three years in jail, less the two years he had already spent in custody.

Argyle does not know how much pink rough went out of the mine and overseas. One account suggested $30 million worth of goods disappeared. The question remained—how many people participated in the conspiracy to rob Argyle? A secret report to Western Australia's police commissioner is said to have identified forty individuals in Western Australia and about thirty elsewhere in the country who might have aided the thefts.

◆

Stolen diamonds move easily into the world of legitimate diamonds. This traffic obeys a single law, the law of demand. The ease of migration from the dark into the light is in the nature of the diamond trade, with its private language and its caste of initiates and its historical origins in small and secretive transactions. Even today many deals are closed in cash. Paper records may be absent. Important dealers who buy hundreds of millions of dollars a year in rough through legal channels may also purchase goods that have no provenance at all except that they have appeared in the office and are for sale. This market distinguishes diamonds from other contraband. Illegal drugs, by contrast, must be distributed by criminals, because criminal channels are needed to move them. The drugs remain in an outlaw world; in many countries the simple possession of them is a crime. But a stolen diamond swiftly exits the criminal world. The moment it passes from a thief's or smuggler's hand into the hand of someone licensed to deal in it, it acquires a pedigree. Now it is simply stock-in-trade, indistinguishable from rough that was bought from legitimate sources. One of the most breathtaking frauds in diamond history relied upon this anonymity of the goods.

In 1992, a slim, immaculate Russian in his early thirties breezed through London's Heathrow airport with the ease that comes with a first-class ticket. Andrei Kozlenok boarded a flight for San Francisco. On arrival in California, he quickly cleared customs and immigration. The picture of a prosperous young executive, Kozlenok got into a waiting limousine and rode into town. He wore an Italian suit and a $50,000 watch, and he had with him a most amazing business plan: to import Russian rough into the United States, polish it in San Francisco, and sell it into the American market.

Kozlenok framed his remarkable proposition as a direct assault on the De Beers cartel. The agreement under which De Beers bought Russia's rough would expire in two years, and the Russian government, always suspicious of the cartel's pricing regime, was scouting for alternate ways to sell its goods. Kozlenok offered such an alternative—bring the rough straight into the world's biggest diamond market and manufacture the polished goods there. In San Francisco, Kozlenok formed a partnership with a pair of immigrant Armenians, David and Ashot Shagirian. The three men took the first initials of Andrei, David, and Ashot, and named their company Golden ADA.

Golden ADA was, from first to last, a gloriously messy cataract of events, and even though it has been widely aired, including such frontline reportage as a cover story in *U.S. News & World Report*, it retains in the diamond trade something of the quality of an unwelcome family secret. Kozlenok's little company had connections at the highest levels of the Russian state. Its contacts reached all the way to the Russian president, Boris Yeltsin. One such contact was Yevgeni Bychkov, who in 1985 had been appointed head of the State Treasury. As part of its responsibilities, the Treasury superintended the underground vaults, hidden throughout the country, that contained the national treasure of Russia. There were fantastic inventories of art and furniture, precious objects, jewels such as emeralds and rubies, silver, platinum, and about 140 tons of gold. There were also diamonds. In one repository in Moscow, thirty feet beneath the street, lay an entire catacomb stuffed with bags of high-end rough. Russian civil servants referred to these secret underground warehouses collectively as "the closet."

In 1992, Boris Yeltsin expanded Bychkov's duties to include the chairmanship of the Committee on Precious Metals and Gems. One of Bychkov's new tasks would be to explore new marketing channels for Russian rough. Bychkov had already distinguished himself in the diamond sector, according to one report, by losing $22 million that the Russian government had expected to gain from a 1990 secret diamond sale. Bychkov proposed to his superiors that Russia open its own direct diamond pipeline to the United States. He would start by shipping goods to Golden ADA in San Francisco. Golden ADA would use the goods as collateral to raise a loan from an American bank. With the loan, it would establish a Russian diamond-polishing and sales office in San Francisco.

A huge load of treasure, worth some $90 million, arrived in San Francisco. There were not only diamonds but tons of gold coins dating from czarist Russia. A bullion dealer in Los Angeles sold coins worth $50 million; the proceeds went to Golden ADA. Kozlenok began to live up to his income. In one day he bought a Rolls-Royce and a pair of Aston-Martins, spending more than $1 million. The rush of revenue into Golden ADA also produced a little fleet of pleasure boats costing more-than $1.2 million. There were more cars too, so that the company ended up with a total of fifteen vehicles.

According to those who later examined the books, Golden ADA bought an $18 million Gulfstream jet, condominium residences worth $3.8 million, a chain of gas stations, and the $4.4 million Lake Tahoe estate where *The Godfather, Part II* had been shot. Kozlenok built himself a $5 million house in San Francisco and decorated it with expensive paintings, Fabergé eggs, and a pair of five-foot clocks in gold cases. Golden ADA made friends among the city's politicians. The three principals posed for a picture with Al Gore, then vice president of the United States. They donated a Russian military helicopter to the San Francisco police, with the sole proviso that Golden ADA could use it to ferry bags of rough from the airport to their offices downtown.

The rough flowed in. At one point Russian officials shipped some 90,000 carats. The terms of the ostensible agreement between the Russian state and Golden ADA called for the rough to be polished in Kozlenok's factory. He did construct a factory, in the San Francisco building that Golden ADA had bought for cash. It had modern equipment and a staff of thirty-six diamond cutters. To anyone in the diamond business, those two facts together—90,000 carats and thirty-six cutters—would raise a flag of alarm. A 1-carat piece of rough can take two weeks to transit a factory. In that time it would pass from sawing to bruiting to the polishing of the crown facets and pavilion facets and girdle facets. One cutter would do one job on a stone, and another would do the next. The polishing of the hypothetical 1-carat rough diamond into a half-carat polished diamond might easily consume four man-hours. With 36 polishers working a 40-hour week, the factory had at its disposal 1,440 man-hours a week, enough to process 360 carats of rough. At that rate it would take 250 weeks, or almost five years, to get through 90,000 carats. The arrangement did not make sense. What was happening to the rough?

In late 1994, a file with information on Golden ADA was dropped on the desk of a thirty-nine-year-old agent of the Federal Bureau of Investigation (FBI) in San Francisco. Joe Davidson had worked on such investigative targets as the Colombian drug cartel and the American Mafia, and the new file caught his attention right away. FBI informants were reporting that Kozlenok and the Armenians were spending huge sums. Only a few months before, Kozlenok had hosted a gala evening on the roof of Golden ADA's building, at which the helicopter gift to the San Francisco police was parked on its helipad for the guests to admire. San Francisco's mayor and chief of police, influential California state bureaucrats, business executives—all milled around on the roof eating caviar and drinking champagne and listening to Kozlenok tell them how San Francisco was going to be a world diamond center.

The FBI began its investigation by establishing contacts within the Russian interior ministry, which controls the police. The Americans formed a working group with the Russians. A member of that group was Major Viktor Zhirov, a veteran investigator with the interior ministry's financial crimes unit. Zhirov had already been tipped to the existence of Golden ADA by a query from American customs officers. The Americans had asked Interpol, the international police organization, whether certain large shipments of diamonds and gold into the United States were legal, and Interpol had routed the query to Zhirov. Zhirov attempted to examine files at the ministry of finance in Moscow, since the finance minister's signature had been required to authorize the removal of assets from the state vaults. The policeman found the files he wanted had been classified, as were the relevant customs declarations and correspondence. But Golden ADA maintained a Moscow office, so Zhirov raided that instead, seizing documents that described the export of $88 million in rough. He passed this information to the FBI's liaison man in Moscow, who fed it back to San Francisco.

Kozlenok learned that the FBI had opened an investigation of Golden ADA. He turned for help to Jack Immendorf, a senior adviser to San Francisco's mayor, offering Immendorf a salary of $1 million to take over as chief executive of Golden ADA. Immendorf, a seasoned private investigator, agreed, and in turn hired Quentin Kopp, a California state senator, as Golden ADA's counsel. They found the company in a chaotic state, but with amazing assets—literally treasure!—ready to hand. Money was pouring out of the company's accounts. Immendorf brought in the accounting firm of Arthur Andersen to examine the books. They found that Golden ADA had paid out $130 million, and they could not tell why. Immendorf quit in frustration after six months on the job.

Soon after Kozlenok's flight, Zhirov arrived in San Francisco. At first the FBI were reluctant to trust him. Russian organized crime was known to have corrupted Russian police, and the FBI were wary of providing information to someone who might possibly leak it to criminals. But as Joe Davidson got to know the Russian, he became convinced of his honesty. For one thing, Zhirov wore the same suit every day; it seemed to be the only one he owned. Then too, he ate inexpensive snacks for lunch and dinner so he could pocket the per diem allowance the FBI paid for food. Davidson discovered that Zhirov made $400 a month; the American per diem was a bonanza for him. He was obviously a man who had only his salary to depend on. The more Davidson saw of Zhirov the more he trusted him, and finally, the more he worried about him. Zhirov was nosing around among the affairs of powerful Russians. One of his subordinates back in Moscow was threatened with a knife in the subway, and warned to keep away from Golden ADA. Zhirov responded by increasing the pace of his investigation.

When he returned to Moscow, Zhirov put more agents on the case. A few days later, two men ambushed him and beat him badly. "Stop the Golden ADA investigation," one of them shouted in his face when the attack was over, "or next time we'll kill you." The assault put Zhirov in the hospital with a concussion. When Davidson heard about the attack, he thought it meant the end of the investigation. Instead, Zhirov left the hospital and went straight back to the case, putting even more police into the effort. Yevgeni Bychkov, the chairman of the state Committee on Precious Metals and Gems, was now being asked what the state had received in return for $178 million of its treasure. Bychkov blamed his former protégé, Kozlenok, and in late September, only weeks after Kozlenok had fled the United States, the Russian treasury launched a suit against Golden ADA in the United States federal court in San Francisco, charging the company with theft of Russian treasure.

As the suit opened, the FBI pressed on with its investigation. It produced a fantastic tangle of leads that pointed to corruption at the highest levels of Russian government, including the office of President Boris Yeltsin. A year after the lawsuit in San Francisco began, the United States Internal Revenue Service (IRS), fearing that Golden ADA's assets would evaporate before the litigation ran its course, sent fifty agents ranging across northern California in a raid. They seized gas stations. They intercepted a shipment of gold, jewelry, and rough diamonds headed for Switzerland. Of $178 million in Russian diamonds and other treasure shipped out of Russia, the IRS recovered $40 million. The agency kept $10.5 million, gave $25 million back to the Russians, and distributed the rest among the numerous, unhappy creditors of Golden ADA.

The Russian government dissolved Bychkov's agency and placed control of the treasure in the hands of the minister of finance. In addition to Bychkov, Russian prosecutors indicted thirteen others. Bychkov was charged with abuse of power but was later released under a general amnesty decreed by Yeltsin to mark the fiftieth anniversary of the end of World War II. Fresh from custody, he assumed the vice presidency of a Russian bank with ties to the diamond industry. Kozlenok disappeared from Belgium, and Zhirov tracked him down in Greece. Russian prosecutors extradited him and returned him to Moscow. On May 17, 2001, a Moscow court sent Kozlenok to prison for six years and levied a $1.8 million fine against him. The same court found Bychkov guily, again, of abuse of power, but because he had in the past received an award of distinction from the state, he was immediately freed. According to an account of the convictions by The Associated Press, the conviction of high Russian government officials on corruption charges is rare, "even though graft is rampant throughout the country."

Back in the United States, Ashot Shagirian settled his affairs with American authorities. As this was being written, David Shagirian, according to the FBI, remained at large.

Investigators concluded that the purpose of Golden ADA had been to obtain not only gold and diamonds but oil from the Russian strategic reserves. It was only the commercial ineptitude of Kozlenok that prevented this from happening, and had it not been for his ostentation and calamitous financial management, he might be running Golden ADA still. Taking into account the small amount handed back to the Russians by the IRS, Kozlenok's thefts had cost his country $153 million.

The essence of this story is that diamonds are not hard to sell. There was never a possibility that Kozlenok's factory in San Francisco would polish all the rough that arrived in bags from Russia. The bulk of the diamonds went straight to the place where Kozlenok knew he could sell them—Antwerp. Belgian customs accepted the origin of the rough as Zaire, and waved it in. In the diamond quarter of Antwerp, the distinctive characteristics of Russian rough would have alerted dealers that the goods were not African; they bought them anyway.

A shadow world of illicit diamonds exists alongside its legal counterpart, and the line between the two is blurred. Diamonds from one side of the law slop across onto the other, and who is to say which are which? Hundreds of millions of dollars in rough are stolen from De Beers–operated mines each year. In turn, De Beers buys rough on the open market, and certainly these purchases include diamonds that were stolen from it in the first place. It is said by some, including the FBI, that Russian organized criminals steal as much as 40 percent of the country's production, although Richard Wake-Walker, who is intimate with the Russian diamond scene, dismisses this figure as too high. However much is stolen, the looting is possible because the retail buyer at the end of the line cannot tell a lawful diamond from an unlawful one. This simple truth permitted a cynical complacency to become seated in the diamond world, and led to the establishment of an infamous trade: war diamonds.

9 The Diamond Wars

In the middle of the 1950s, De Beers discovered that illicit diamond dealers were buying large amounts of rough from the alluvial fields in Sierra Leone and smuggling it out of the country. To stop this trade, Harry Oppenheimer hired Sir Percy Sillitoe, the retired chief of Britain's spy agency, MI-5. Sir Percy assembled a team of mercenaries, named it the Diamond Security Organization, and began to intercept the smugglers' caravans as they made their way through the bush to Liberia. The smuggling stopped, at least for a while.

Half a century later, Sierra Leone is caught in a vicious trade known variously as conflict diamonds, blood diamonds, and war diamonds. Bandit armies rampage through the country in a war for control of the diamond fields. As before, the gems move out through Liberia. Sir Percy would be harder put today to halt this trade. Even crack British troops, supporting the United Nations, suffered a humiliation in late 2000 when some of their soldiers were captured by a teenaged rabble with rifles from Russia and brains ablaze with drugs. But it is the crimes against civilians—maiming and murder—that have attracted the most disgust.

Press coverage of the diamond wars has aimed a floodlight of scrutiny onto the diamond trade, illuminating what was secret. The trail of implication led to Antwerp, Tel Aviv, Bombay, and in Africa involved more than a dozen countries. In 1996 and 1997, at the peak of the African diamond wars, some $1 billion in illicit goods came out of the continent every year. Behind the trade lay ruined cities, hundreds of thousands of dead, and a history that went back a quarter century.

The worst of the diamond wars commenced in 1975 when the Portuguese colonial government of Angola, facing insurrections they could not subdue, abandoned the country. Half a million white colonists fled Angola, which dissolved into civil war. The main antagonists were the Popular Movement for the Liberation of Angola (MPLA), and the National Union for the Total Independence of Angola (UNITA). Cuban troops arrived to help the MPLA; the South African army and air force, with the tacit approval of the United States, sided with UNITA.

UNITA captured the prime Angolan diamond rivers, and the MPLA relied for its revenue on offshore oil. Well supplied with money, the war rolled back and forth across the country. There was a brief hiatus for elections in 1991, which were supposed to end the war and supply the country with a mandated government, but the results were disputed and the war resumed.

By 1993 UNITA controlled almost three quarters of Angola. Its diamond revenues were $600 million a year, giving it the ability to prosecute its war. But in 1994, under pressure from the United States and Russia, the combatants signed a treaty whereby UNITA would disarm in return for assuming a place in government. After twenty years of war, official peace reigned in the ruined country. Although UNITA did not in fact disarm its main battle regiments, the MPLA government in Luanda was able to occupy some of its adversary's old diamond territory, and the word spread in the international diamond-mining community that Angola was open for business.

Many new mining and exploration ventures sprang up in the following years, all of them jointly with the Angolan "government," that is, the United Nations–recognized MPLA regime in Luanda. The scent of top Angola goods went out into the diamond world, and fortune hunters followed the smell. In 1995 a pair of South Africans suctioned a 24-carat pink out of the Chicapa River. They sold it a week later at the diamond bourse in Johannesburg for $4.8 million. The stone went directly to New York, where it was reputedly resold for $10 million. According to the story abroad in the trade, the pink was then polished and sold to the sultan of Brunei's brother, for $20 million, another example of the kind of goods that had kept an army in the field for twenty years.

Two years later, in 1997, Chris Jennings organized a shuttle of Russian-built Ilyushin freighters that flew a relay for seven straight days out of Johannesburg's Jan Smuts airport, carrying mining equipment to the city of Saurimo, and trucking it over dirt roads to the place on the Chicapa where the pink had been found. The very next year another Canadian junior, DiamondWorks, arrived on the Chicapa. This commercial activity was supposed to prove that peace had come to Angola. Nevertheless, there were huge tracts where UNITA's writ still ran and where Jonas Savimbi, the wily general at the head of UNITA, was the undisputed monarch. Because of this, some investors were shy of putting money into Angola, and early in 1998 DiamondWorks invited a group of stock market analysts to visit the Chicapa and see how safe the country had become. And so it was that at 5:30 one January morning, a small group rode out to Johannesburg airport, boarded a chartered Boeing 737, and with champagne and orange juice in good supply, took off for Angola.

(DiamondWorks had been dogged by news accounts—always flatly denied—linking it to the South African mercenary group Executives Outcomes. And such troubles would not go away. In January 2000 Peter Hain, Britain's foreign office minister for Africa, named Antonio Teixeira, a South African mining executive, as one of a number of people dealing in diamonds proscribed by the United Nations. Teixeira, at about thaqt time a newly-appointed director of DiamondWorks, vehemently rejected the truth of Hain's words, and challenged the minister to repeat them outside the legal shelter of the British House of Commons).

The Angolan capital, Luanda, is sited on the shore of a beautiful bay. The Atlantic Ocean sparkles in the sun. At first glance from the air, the city looks as it must have looked when the Portuguese were there and Luanda was a graceful strip of urbanity with high-rise apartment buildings and curving boulevards. But as the plane descended, the city revealed itself as a ruined shell, its buildings and roads broken by neglect and the privations of war. The 737 coasted down onto the runway, slowed, and taxied past lines of battered old Antonov freighters parked in the grass.

The DiamondWorks party cleared customs, reboarded the jet, and an hour and a half later landed at Saurimo. A helicopter waited to carry the group on the last leg, and after a twenty-minute flight the meandering stream of the Chicapa appeared. The diamond ground of Luo lies at a bend in the river. DiamondWorks had diverted the river and was stripping the bank. A small recovery plant chewed away at the gravel. On the far side of the river, trees laden with white blossoms trailed their branches into the muddy water. A barge was tethered in the current and a plastic float marked the place where a diver made his way along the bottom with a suction hose.

After a tour of the Luo site, the helicopter headed downriver to view another DiamondWorks property. On the way lay a sixty-mile stretch of river, traditional UNITA mining ground, and the shore was pocked with signs of recent mining. When the helicopter reached the new mine site, at Yetwene, I tapped a company spokesman on the shoulder and shouted over the noise of the helicopter: "What about UNITA?"

The man shook his head. "They're completely out of the Chicapa."

"Who was that mining back upriver?"

"I don't know. The area's completely pacified."

Ten months later UNITA commandos came out of the bush and assaulted the Yetwene mine. Eight people died in the firefight. The cease-fire crumbled. The opposing armies of the MPLA and UNITA moved out of their positions and resumed the war. DiamondWorks more or less fell apart. The small company had lacked the basic equipment required to mine diamonds in Angola—an army.

◆

In a new model of thinking about the diamond wars, the fighting is viewed as straight commercial activity. Between 1992 and 1999, Angola's civil war killed some five hundred thousand people and supplied UNITA with income from diamond sales of almost $4 billion. Research at the World Bank suggests that in such conflicts the objective of defeating the enemy in battle has been replaced by the profit motive, so that the diamonds are not just the means to prosecute the war but the reason.

Angola's rough is worth an average of $200 a carat, which is high. (The smallest diamonds in any sample pull down the average value; a representative parcel of Angolan rough would contain many stones worth thousands of dollars a carat.) But even in the neighboring Democratic Republic of the Congo, where the average price for rough is only $20 a carat, diamonds have helped attract the soldiers of several armies. Namibia, Zimbabwe, Angola, Uganda, Rwanda, the Congo itself, as well as numerous guerrilla groups with shifting affiliations roam the Congo in a grotesque melee that has been called the First World War of Africa.

When Laurent Kabila, the Congolese president, was assassinated in January 2001, the wars had put some thirty-five thousand foreign troops into his country, killed one hundred thousand of its citizens, and sent 1.3 million refugees into the bush, prey to malaria, meningitis, and starvation. While some of the foreign soldiers were there to prop Kabila up and some to tear him down, all were there to capture assets. Some invading generals seized a diamond mine and formed a corporation to market the goods. In the east of the country, other invaders grabbed a diamond-producing zone of their own.

To disguise the origin of war diamonds, traffickers move them through certain countries that abet the trade, such as Liberia, a colossal diamond laundry through which large amounts of illicit rough are rinsed on the way to Antwerp and Tel Aviv. In 1996, for example, according to the United States Geological Survey, Liberia's production of rough diamonds from its own sources totaled 150,000 carats; yet in that year the country shipped to Belgium 12.3 million carats, or eighty-two times what it had produced. Anyone accepting the Liberian figures at face value would have to believe that Liberia had outproduced the whole of South Africa. Yet the Diamond High Council, the Antwerp diamond lobby entrusted by the Belgian government with monitoring the ebb and flow of diamonds, simply recorded the origin of the goods as Liberia. Whatever else its failings, the Antwerp diamond quarter is not naive. The true origin of the goods was widely known.

◆

The diamond wars were the secret of the diamond trade until, quite suddenly, they were not. It seemed to happen in an instant, as if a curtain had been ripped aside and there was the diamond business, spattered with blood, sorting through the goods. Its accuser was a little-known group called Global Witness. In the jargon of activism, Global Witness was an NGO, or nongovernmental organization. NGOs campaign for everything from famine relief to the abatement of automobile emissions. Global Witness had previously investigated illegal lumbering in Cambodia, but in 1998 turned to examine the Angolan diamond war. In December of that year it produced *A Rough Trade*, a fourteen-page indictment that landed with a thump in the forum of public opinion and caught the diamond trade by surprise.

The booklet attacked the diamond business for colluding in the war. It used statements pulled from various public documents to demonstrate the ease with which Angolan war goods flowed into Antwerp. The United Nations had been struggling against the war trade itself, without much success. *A Rough Trade*'s success lay in presenting the details of the war in an accessible form. Statistics were laid out for easy reading, with maps and chronologies and even a short glossary. A photograph showed a pit holding the mud-caked bodies of victims of the war; another depicted long lines of miners in slave conditions passing buckets out of a pit. A table of figures listed the annual revenue from illicit diamond sales. By leafing through the booklet one could gather at a glance the sordid history of the war, its human and material costs, and, in language made chilling by the context, the commercial considerations of the diamond business.

> That we should have been able to buy some two-thirds of the increased supply from Angola is testimony not only to our financial strength but to the infrastructure and experienced personnel we have in place. [Chairman's statement, De Beers annual report, 1992]

> The CSO buys diamonds in substantial volumes on the open market, both in Africa and in the diamond centers, through its extensive network of buying offices, staffed by young buying officers often working in difficult conditions. Purchases in 1996 reached record levels largely owing to the increased Angolan production. Angolan diamonds tended to be in the categories that are in demand, although in the main these buying activities are a mechanism to support the market. [Chairman's statement, De Beers annual report, 1996]

Global Witness helped to move such information from the general knowledge of the diamond trade to the general knowledge of anyone who picked up a newspaper. The press seized on the disclosures. The diamond world reacted by trying to guess who might be behind Global Witness. Amazing exchanges of e-mails went to and fro, with the most rococo, and plainly absurd, explanations for the group's exposure of the war trade: Global Witness was in the pay of the Angolan offshore oil companies, who wanted to ingratiate themselves with the MPLA government in Luanda by wrecking UNITA's diamond business; De Beers itself had funded the report in order to take UNITA's diamonds off the market, so they wouldn't have to buy them; the United States government controlled Global Witness through the Central Intelligence Agency; BHP had set the whole thing up to cast a shadow on African goods, thereby creating a warmer market for Canadian diamonds. Completely useless as a guide to anything but the Byzantine imagination of the diamond world, the rumors drifted away.

If any doubts remained about Global Witness being suborned by riches, the idea would not have survived a visit to the NGO's office—a cramped pair of rooms on the top floor of a former schoolhouse on a drab north London street. The staff worked amidst a clutter of papers, which they hurriedly turned facedown on their desks when a visitor came by. Charmian Gooch, the head of Global Witness, is a tall, pale woman with an air of intense conviction. From 1989 to 1993 she worked for the Environmental Investigation Agency, a London-based NGO. She started Global Witness in 1993 to research the connection between natural-resources companies and human-rights abuse, starting with Cambodian forests and moving to diamonds. Alex Yearsley, a researcher, had previously been a host for a music video channel. They seemed an unlikely pair to have sent such discomfiture into a $6 billion-a-year trade.

Less than a year after the release of *A Rough Trade*, Global Witness aimed another punch at diamonds. With a talent for marketing that might have impressed De Beers, Gooch and Yearsley packed cut-glass "diamond" rings into black velvet boxes, and sent them off to editors around the world. The package included literature that called the diamond trade a "lethal dinosaur," and included tallies of the dead from the diamond wars. *Fatal Transactions*, as this campaign was called, had powerful allies. Robert Fowler, Canada's UN ambassador, had seized upon Canada's tenure of the presidency of the Security Council to step up his castigation of the diamond trade in general for its systematic flouting of UN resolutions. In Washington, Ohio congressman Tony Hall sponsored a measure aimed at requiring American importers of rough and polished goods to demand certificates of origin from their suppliers.

Opposition to the traffic increased. The British government, embarrassed by revelations that it had countenanced the sale of arms to mercenaries in Sierra Leone, made a grant to Global Witness, and vowed to help end the trade in war diamonds. De Beers caught the way the wind was blowing, and announced it would immediately close its buying office in Angola and embargo the purchase of Angolan rough on the open market. "De Beers," said the company, "shares the world's concern over the continued suffering of the people of Angola, for thirty years the victims of a brutal and devastating civil war." This protestation met with skepticism. A columnist wrote in the Johannesburg *Star* that it took the involvement of the United Nations, the United States Congress, the British government, and Global Witness to prompt De Beers to close its office in Angola.

As publicity spread about the diamond wars, one nagging question took shape: What if the trade in war diamonds could not be blocked? It was a chilling question, raising the specter that any diamond could be a war diamond. After all, the volume of the goods was immense. In the corrupt milieu of Angola, moreover, generals in the government army were widely believed to trade in UNITA rough. If UNITA could sell diamonds to its own enemies, how much more easily could it, and any other army with diamonds to sell, infiltrate the goods into the legitimate diamond stream? This was a fair question, given the sophistication of the diamond smuggling operations.

In a typical scenario, a Russian crew flies an Ilyushin IL-76 transport to Sofia, the Bulgarian capital. The IL-76 is a four-engine jet with a payload of fifty-two tons and a range of three thousand miles. In Sofia the crew picks up the merchandise ordered by UNITA—say, a Russian T-62 main battle tank. Such equipment is not hard to get: you can buy it on the Internet. Ground staff in Sofia prepare a manifest listing the cargo as machine parts. The manifest is tucked into an envelope of cash and handed to Bulgarian customs. The Russians drive the tank into the cargo hold of the jet.

The flight plan lists Lusaka, Zambia, as the final destination. The crew flies to Uganda, refuels, then resumes the flight in the direction of Zambia. They reach Zambian airspace at nightfall, where, under cover of darkness, they abruptly change course to the west and fly to Angola. At night the plane is invisible to American satellites, which could otherwise track its course and report it to the Angolan military.

Once inside Angola, the Russian pilots head for a landing site specified in advance by UNITA. The landing strip might be listed on maps as deserted, but in fact be maintained by UNITA. Or it might be a new strip—a straight stretch of road bulldozed flat and watered. The wet dirt hardens in the sun. UNITA never widens such a strip, because a widened road is visible from space and will be detected by a computer tasked to scan satellite images for just such a change. Because the roads are narrow, and the bush on either side would tear the wings off a plane, UNITA prunes the vegetation beside the road to a height of two feet. From space, the trimmed vegetation is indistinguishable from the taller vegetation around it, yet low enough for the Ilyushin's wings and engines to clear. As with all tactical military transports, the Ilyushin's wings are level with the top of the fuselage, which helps to keep the jets from sucking in dirt and other loose material.

At prearranged coordinates the pilots begin a rapid descent, and at an altitude of a thousand feet transmit a coded signal. A UNITA crew turns on a line of landing lights, or diesel-burning torches, hooded against detection from space and aimed in the direction of the jet's approach. At 140 miles an hour the huge transport comes in over the bush and thunders onto the dirt. The pilots apply reverse thrust and the plane screams to a stop. The transaction itself is simple, with well-established protocols. The diamonds to be exchanged for the tank have already been checked by a broker resident in UNITA territory. Now they are checked again on board, the tank rumbles out, and the plane takes off with the diamonds.

Such rough would move easily into the mainstream of legal goods. For example, the flight might stop in Guinea, where the rough would be washed through some diamond-trading house authorized to attach a certificate of provenance. Or the crew could take the diamonds to Lusaka, mix them with other African rough, and bring the goods to Kiev, the Ukranian capital. Factories in the Ukraine would then polish the diamonds alongside Russian rough. Once a diamond has been polished, its country of origin is virtually impossible to tell. Or instead of the Ukraine, the shipment could go to Russia to be mixed in with Russian rough and sold into Antwerp. It is hard to tell the origin of rough that has been mixed with other rough.

(A competent diamantaire can tell the origin of a sample of rough if he has a parcel with goods from only one source, so that distinguishing characteristics such as color, shape, and size are readily apparent. But the clear picture presented by such attributes is muddied and becomes unreadable when alien rough is mixed into the parcel, introducing a range of particularities different from those of the original goods.)

A favorite transshipment point for illicit rough was Liberia, where diamond companies from Antwerp and Tel Aviv had buying offices. In the Liberian diamond laundry, war goods and goods stolen from mines came together. Western intelligence agencies believed Liberia to be the center of an international criminal consortium founded on a trade in drugs and diamonds. Liberia's links to war diamonds were exhaustively detailed in January 2000 by Partnership Africa Canada, an NGO whose researchers came closer than previous critics to advocating the harshest possible response to the diamond trade's entanglement with war: a consumer boycott of all diamonds.

◆

At this point, early in 2000, the trade was in disarray. The barrage of press reports intensified. The oft-heard defense—that blood diamonds accounted for less than 5 percent of the whole trade—sounded grotesquely ignoble and false. De Beers had already perceived that the calamity of a consumer boycott was a possibility, and had moved to distance itself from contaminated goods. The Diamond Trading Company's offices in Antwerp had ceased all purchases of open-market rough, and had even stopped buying the "official" Angolan production that came out of Luanda. In March 2000, De Beers began inserting into sight boxes notices guaranteeing that none of the goods had been purchased in contravention of UN resolutions against the war trade. It was a seminal event, serving notice that the leading diamond house was setting itself against the bloodied goods and aligning itself, in effect, with some of the harshest critics of the trade. Some diamond insiders questioned the reliability of De Beers's notices. The London stockpile, they grumbled, must certainly contain goods purchased in former times, when the DTC was a vigorous open-market buyer. De Beers dened such allegations, insisting that internal audits allowed them to assert that the goods were squeaky clean. That doubts persisted anyway shows how completely the broader trade had been compromised by illicit goods, to the point where knowledgeable observers could not believe that the mass of goods available in such centers as Antwerp was not hopelessly contaminated. Yet even to such doubters it was plain that De Beers had decided not to follow events, but to lead them.

Only De Beers could have galvanized the trade into action. Diamond dealers are notoriously fixed on the short term: their business consists of rapidly turning over large volumes of diamonds, often at narrow margins, and this regime focuses attention on the transaction of the moment. A kind of willing blindness screened the trade from the larger issue of morality. There was one notable exception—the diamantaire and publisher Martin Rapaport. In a series of anguished pieces in his monthly magazine *Rapaport Diamond Report*, Rapaport inveighed against the horrors of the diamond wars. Yet nothing sounds so loudly in the diamond trade as the voice of De Beers. In June 2000, Nicky Oppenheimer made clear how acute the matter was when he sent a letter to the diamond bourses urging them to expel any member caught trading in war goods. Oppenheimer bluntly warned that failure to stop the war trade "immediately and willingly, could have incalculable repercussions on the diamond market. . . ."

The following month, at a congress in Antwerp, the International Diamond Manufacturers Association and the World Federation of Diamond Bourses devoted the whole agenda to an emergency discussion of the war-goods issue. UN Ambassador Fowler appeared, and so did high officials of the British foreign office. The congress produced a proposal calling for a system of export and import controls designed to exclude war goods from the trade. Ideally, the system would affix a ticket to any parcel of goods coming out of a mine, so that the parcel could be tracked from mine to market. Legitimate goods would possess a "chain of warranties," as the trade began to call it. Any dealer caught with diamonds not protected by the chain would be expelled from the diamond bourses.

There were doubts about the effectiveness of such measures. Only the month before, a British reporter had brought rough diamonds from Sierra Leone to London's Hatton Garden diamond quarter and found eager buyers. At the time, British law prohibited such goods, and the London diamond bourse had a stated policy to expel any member caught dealing in contraband. The reporter had made it clear to the buyers that the diamonds came from Sierra Leone, and did not disguise their origin as war goods. Some dealers told him they had bought such rough before. The reporter said he could get more, and the dealers expressed an interest. Still, at the Antwerp congress, Ambassador Fowler hailed the new proposals as a sign of the diamond trade's willingness to come to grips with the issue. The diamond manufacturers and the bourses set up a third body, the World Diamond Council, to monitor such tracking measures as would be put in place.

When the congress finished and the last avowal drifted off across the River Scheldt, critics of the trade might well have wondered what had been accomplished. While governments thought that the diamond trade could stop the flow of war goods where the governments had failed, the trade seemed to envision the reverse—a hedge of laws enacted by the governments. But laws and resolutions already existed to interdict war diamonds, and the trade in them had not been much perturbed. In this context, some believe the surest barrier against war diamonds is technology.

In Canada, where the government is anxious to protect its nascent diamond industry from the scandal of the war trade, the federal police are studying a method for determining the point of origin of any rough diamond. The system works by vaporizing with a laser a tiny piece of the rough diamond under inspection. The vapor is then analyzed by mass spectrometry. Although diamonds are considered pure carbon, they do contain small amounts of impurities. In terms of weight, the impurities would amount to less than one twentieth of 1 percent. As many as fifty different elements make up these impurities, and the elements exist in ratios that differ from source to source. A diamond from one pipe will contain a ratio of impurities distinct from the ratio present in diamonds from another pipe. By analyzing the vaporized bit of rough from a given pipe or alluvial deposit, a researcher can retrieve the unique profile of impurities, and can therefore identify the source of the goods. An inspector supported by this technology could pluck a single piece of rough from any parcel and determine its origin. Anyone mixing war goods into a parcel would have to accept the risk of detection.

The difficulty with this system lies in collecting data. A reliable inspection lab would need samples of every diamond population in the world in order to extrapolate reliable profiles. UNITA would be unlikely to supply such a sample, nor would the rebels in Sierra Leone. Or such samples, even if obtained, could be fraudulent. Angolans, for example, could mix UNITA rough into a sample in order to furnish a false profile, which would then allow them to continue to broker UNITA goods.

Moreover, legitimate mining companies would probably resist furnishing adequate samples of their rough because the full range of goods produced by a mine is a commercial secret. Diamond mines have distinct productions of gems. The distribution of diamond sizes and qualities determines the commercial strength of a mine. By discovering that makeup, a competitor would learn which types of goods constitute his opponent's strength, and could flood the market with those goods, attacking his opponent's most profitable business, even if only for a short time. Normal habits of commercial prudence would also militate against the sharing of such information.

Another way to detect war rough, perhaps more promising, would be to analyze the tiny particles of soil that adhere to rough even after it has been cleaned in acid. Research into such a method was proposed by George Rossman, professor of mineralogy at the California Institute of Technology. Rossman made his proposal in response to a request for submissions from the White House Office of Science and Technology Policy, which in January 2001 convened a special conference to look for ways to identify the local origins of rough.

Rossman obtained a sample of Orapa rough from the open market. Although such goods would be acid-washed, he did not think the cleaning process would remove all soil traces. Some minerals, for example, take months to clean. When Rossman examined the Orapa diamond, he found microscopic specks of soil lodged in tiny creases in the crystal.

The specks were remnants of slimy coatings (made up of minerals deposited from local water) that form on the surfaces of diamonds in the ground. If a diamond travels a long way from its source, as alluvials do, a new film of minerals will reform on the surface wherever the stone comes to rest. This new coating reflects the local soil chemistry and, crucially for Rossman's scheme, the isotopic composition of the local water.

Isotopes differentiate otherwise identical chemical elements. The atoms of a particular chemical element may differ from one another in weight because of different amounts of material in the nucleus of each. Such distinct atoms of the same chemical element are called isotopes. Water from any specific place on Earth will have a specific isotopic composition, or signature. These signatures consist of very small but measurable differences in the ratio of heavy water to normal water. Water is made of oxygen and hydrogen; the hydrogen in heavy water has extra matter in the nucleus.

Rossman put the Orapa rough into a furnace and captured the water vapor that came off—water that had been trapped in the minute residue of soil secreted on the surface of the gem. From the vapor, the scientist retrieved 0.0000001 grams of hydrogen, an almost incomprehensibly small sample, yet enough to establish the isotopic signature of the water. The signature obtained matched the measurement indicated for the Orapa region on a world map of hydrogen isotopic compositions. Other source locations could be similarly established; indeed, Rossman had detected tiny amounts of foreign matter on the surfaces of cleaned rough from Angola and Sierra Leone. His proposal called for further research, including the collection of soil samples from conflict areas.

While techniques for identifying the origin of rough are the surest way to choke off conflict diamonds, the polished side of the trade offers its own opportunities. The government of the Northwest Territories is attempting to establish a clear "audit trail" from mine to cutter, channeling goods directly from the Barrens to the new polishing factories in Yellowknife. Sealed packets of diamonds from the mines are opened in Yellowknife, and each rough diamond in the parcel entered into a database that records its weight, color, clarity, crystal type, expected yield, expected color after polishing, and expected clarity after polishing. When the manufacturing process is complete, the finished diamond is matched against the database entry for the original rough. Polishers participating in the government-run program may then laser a polar bear logo onto a girdle facet and supply a certificate warranting that the polished diamond is a genuine Canadian Arctic Diamond.

Of course, anyone can laser a polar bear onto a diamond and, sadly, unauthorized bears are said to have been spotted shuffling out of some unpolar places and into the mainstream of polished goods. To protect its program from counterfeit bears, the government added a security feature called Gemprint, developed by the Gemprint Corporation of Toronto. Generally, a polished D flawless diamond from Sierra Leone, for example, will look the same as a polished D flawless diamond from the Barrens. But not to the Gemprint technology, which "fingerprints" a polished diamond by bombarding the top facet with laser light and recording the unique pattern of pinpoints of light reflected back. The pattern is displayed on the certificate that accompanies each certified diamond, and also digitally stored in Gemprint's database. The spray of dots supplies a unique fingerprint, and an audit trail from mine to consumer is thereby complete. The only place that war diamonds could be inserted into this system in quantities that would make economic sense would be at the mine, an impractical plan but a charming idea, calling up the picture of criminals smuggling diamonds *into* a mine.

The refocusing of attention onto the polished side of the diamond industry by forces that had traditionally stayed in rough is one of the profoundest changes in the diamond world. Even before the issue of war diamonds broke into the public realm, De Beers had begun to test-market the idea of "branded" diamonds—diamonds identified specifically as De Beers diamonds. In the past, diamonds had always been anonymous. It was as if the best car dealers in the world all sold a product known simply as "luxury car." For its $200 million a year in advertising, De Beers enhanced not only its own goods but everyone else's.

In 1996 the De Beers research establishment at Maidenhead, near London, began to investigate ways to put a brand on polished goods. They developed a technology for inscribing a logo onto a finished diamond. The mark is invisible to the naked eye, and the system for inscribing it is a trade secret. A prospective buyer can view the mark and his diamond's serial number through a special device that will be sold to the retail trade by De

Beers. According to the company, a test market in the English city of Manchester in 1998 showed that buyers would pay a premium for De Beers branded goods, and that retailers with such branded goods on offer sold more diamond jewelry overall.

Although this claim is disputed by some in the diamond trade, the branding initiative gained strength from a separate phenomenon—the war-diamonds issue. Gary Ralfe, the managing director of De Beers, freely acknowledges that he instructed De Beers technical staff to speed up their work on marking goods when publicity about the war trade grew. De Beers could envision a line of goods with a war-free pedigree, set apart from other diamonds by the inscription of De Beers's logo on a polished stone. The company already possessed the largest supply of legitimate goods, protected from contamination by the cessation of outside purchases. If it could extend a presumption of legitimacy down into the retail trade, the cartel would gain a marketing opportunity. As the diamond writer and oracle Chaim Even-Zohar put it, "The brave man would write that the whole issue of war diamonds can only benefit De Beers."

Because De Beers is powerful and secretive, and has dominated its industry for so long, the company's motives are often impugned. It should be noted, though, that De Beers has threatened any customer caught dealing in war diamonds with the harshest sanction imaginable—expulsion from the London sights.

◆

The diamond wars added urgency to changes already afoot in the diamond business. The discoveries in the Barrens threatened the old cartel by promising an alternative supply of good gems. It had been apparent from the earliest days of the northern diamond strike that the new producers were contemplating the promotion of their goods as distinct from the cartel, hoping to create, especially in the United States, an appetite for goods that came from outside the contrived, monopolistic system of the cartel, a system that might be made more offensive to Americans simply by the fact of an alternative appearing on the scene.

While these larger struggles played out, the daily business of the diamond trade went on, in some respects indifferent to the forces shaking its foundations. At its heart the trade is absorbed in a world with only a single preoccupation—the stone. In this sphere there is no manipulation of the price of rough, no slaughter of innocents, but only the careful business of releasing light. It is in this separate land that the diamond cutter engages in a hand-to-hand combat with the stone, and raises the diamond world above its sometimes sordid circumstances, moving it onto the plane of the jewel, where it takes its meaning.

10 Diamond Cutter

On a summer's day in 1998, as Manhattan lay baking in the worst heat wave in more than a century, the diamond broker David Danziger walked through the oven of the streets to pay a visit to William Goldberg Diamond Corporation on Fifth Avenue. Danziger rode the elevator up to Goldberg's offices. The receptionist, behind a plate of bulletproof glass, buzzed him in. He entered a tiny, wedge-shaped room, paneled in bird's-eye maple. The lens of a camera pointed at Danziger. When the first door clicked shut, the receptionist buzzed Danziger through into a paneled corridor carpeted wall-to-wall in slate gray broadloom.

The closed doors of four salesrooms stood along the narrow corridor. The receptionist unlocked room number three, and Danziger went in and sat down at a narrow table with chairs on either side, a diamond scale, fluorescent lamps, and blocks of white paper eighteen inches square. Diamond dealers like to study diamonds under intense light against white paper, the better to assess a stone's color.

Danziger waited on a black leather chair. Jewelry catalogs from Sotheby's and Christie's lay neatly stacked on the broad windowsill. The muted sounds of Manhattan traffic reinforced the cocoon of carpet and paneling. Beyond the window stood a cedar in a wooden tub, at the very end of the terrace that wrapped around two sides of William Goldberg's private corner office. On the Fifth Avenue side the terrace was wide enough for tables and chairs, and on pleasant days the Goldbergs' chef served lunch outside.

Danziger did not have to wait more than a few minutes before he heard William Goldberg's husky voice coming along the hall. One of the great diamantaires of New York, Goldberg is tall and powerfully built, and he creaks through his domain like an old lion, guarding neither his opinions nor his voice. His long white hair hangs over his collar. The top of his head is a smooth, bald dome, tanned to the color of dark wood. His eyes are warm brown, and twinkle beneath extravagant eyebrows. He opened the door of salesroom number three and waved Danziger back into his chair and shook his hand firmly. The broker wiped his forehead with a handkerchief. They agreed the heat was cruel, and without more ado Danziger fished in his pocket and produced a folded packet of white paper. "I know this isn't your type of goods, Bill," he said, putting the packet on Goldberg's desk. "Probably you won't want it." He shrugged. "I thought you might like to take a look anyway."

A penciled notation on the packet detailed the contents, so when Goldberg flicked the paper open with his thumbs and tumbled out the stone, he already knew it weighed 82 carats. It was not a pretty piece of rough. It had sharp edges and a jagged shape, as if the steel crushers that chew a mine's ore into manageable chunks had broken this piece from a larger crystal. It had many specks and flaws. The color was not good, a cape yellow, not yel-

William Goldberg. (William Goldberg Diamond Corporation)

low enough to be a true fancy color. And there was something else against it—the way it came in the door.

As a De Beers sightholder, Goldberg's goods would come mainly from his regular London box. Added to these were purchases in Antwerp or Tel Aviv, or even in Brazil if news of some great stone happened to take him there. But a stone arriving in Manhattan in the pocket of a broker is a different matter. It has already passed through many hands. Dealers like Goldberg feel that such a diamond has lost some of its illusion. The magic has been pawed away by other diamantaires. Too many eyes have peered at it through loupes, and found something wrong. Many dealers would have just shrugged at Danziger and spread their hands.

"But Bill isn't too concerned what other people think," said Barry Berg, Goldberg's son-in-law and partner. "He brought it in and tossed it on my desk and asked me what I thought. Well, I thought it was a big, bluffy piece of rough that looked like we could get a nice big polished diamond out of it. I'm not saying the yield would be great. We would get maybe a twenty-carat polished out of it. But we didn't look too closely. Bill went right back to Danziger and offered on it, and got it. He's like that, he shoots from the hip. He wanted to create some excitement in the office. We hadn't had a big stone in a long time, so he went for it." Goldberg paid Danziger $60,000, a sum that reflected the fact that the diamond was riddled with flaws. They put the stone away for the weekend.

On Monday morning four inquisitors clustered around a bench in Goldberg's little factory, in a corner of the fourteenth floor. The cape stone lay in its opened packet in the powerful light of Motti Bernstein's lamp. As well as Bernstein, a cutter, there were Goldberg, Berg, and Ben Green, the cleaver. With the spread of high-speed saws and lasers, cleaving is a dying craft, although it is still preferred by some as the cleanest, fastest opening move in the work of extracting a polished jewel from a piece of rough.

Diamond crystals grow by adding layers. These layers align themselves in planes, which diamond cutters call the grain of the stone. Cleaving splits the stone along a plane, thus the "cleavage plane." The cleaver begins by rubbing another diamond against the diamond to be split, until he has a notch in which to place the cleaving knife. He gives the knife a sharp rap and either gets the two pieces that he wants or destroys the stone. The tale is told of Joseph Asscher, the greatest cleaver of his day, that when he prepared to cleave the largest diamond ever known, the 3,106-carat Cullinan, he had a doctor and nurse standing by, and when he finally struck the diamond and it broke perfectly in two, he fainted dead away. Asscher's nephew snorted when he heard this story. "No Asscher would faint over an operation on a diamond," he snapped. "He's much more likely to open a bottle of champagne."

The party that had gathered around the big yellow diamond at William Goldberg's was trying to decide whether a certain protrusion should be ground away or cleaved. They discussed the stone for fifteen minutes, then agreed to suspend deliberations until the next day. This would give Bernstein time to polish in a window. Through the window they would be able to look straight into the diamond. By positioning the window to inspect

Motti Bernstein marking the yellow. (Matthew Hart)

the major flaw, the cutters would be able to look right down along it as it snaked jaggedly into the crystal.

The decision about cleaving was not their only concern. On closer inspection Berg had thought the stone looked "sleepy," or somewhat clouded, which could mean that there were other flaws, too tiny to be seen individually but numerous enough to cloud the diamond. He could not be sure. The cutter's task would be to retrieve a hypothetical jewel, disguised by the enclosing rough. In this case, he would have to detect this notional stone through the refractive chaos that surrounded it, because the yellow was a mess. It was as if they were trying to judge the shape of a nugget of glass encased in a cracked cube of ice. Motti Bernstein put in the window and had a look and told William Goldberg that it opened well, that the yellow looked better than expected.

Next day Berg convened a meeting in one of the salesrooms. Bernstein came in, plucked a candy from a dish, and deftly removed its golden wrapper, his thick fingers flicking the paper aside like an empty diamond packet. He popped the candy in his mouth and sucked on it speculatively. The 82-carat stone sat on a white-paper pad. "This one," he said, tapping the diamond, "we know it's not a clear stone. So I have to open it up to plan the stone. First of all, we thought maybe it had a lot of fluorescence. Maybe it's milky. But I have a look and it's pretty clear. It has fluorescence, but not so bad. Next, it has some flaws. I thought they might be quite deep, maybe worse than we thought."

A knock sounded at the door and Berg opened it. In came Ben Green, sweating from the savage heat outside. Green was scrupulously formal, shaking hands all around before he took his place beside Bernstein. With one hand he removed his fedora and with the other he put on a skullcap. Then he gazed at the stone with calm green eyes. Bernstein said, "You didn't see it since I opened it."

Green picked it up. "The quality is not so bad."

"No surprises," Bernstein said.

"No surprises," Green repeated.

"It's not clouded."

"There is light there."

"Let's keep the table where I put it. This gletz here doesn't bother me. I don't see it shooting in too deep."

Green put a fingernail against the stone. "There's a knot too. That doesn't bother me."

"You can cleave a knot," said Berg.

Green nodded. "Yes, and it's impossible to saw along it."

A knot is a place inside the crystal where the structure alters. Possibly a larger diamond has engulfed a smaller one. The result is a different orientation of the plane. Diamond cutters liken it to a knot in wood. *Gletz* is of Dutch origin, and means a fracture. The table is the largest polished surface of a jewel—the top. The jargon of the trade went back and forth. They might have been doctors discussing a difficult but interesting patient. The diamond was picked up and put back down. Green drew a line in ink around the base of the crooked protrusion. He said, "People think with cleaving you take a hammer, like in the movies."

Bernstein made a loud sucking noise on his candy and wagged a finger. "Not," he said firmly.

"Not," assented Green, raising a finger himself and moving it back and forth. "Only a small tap with a knife. It is all patience. A cleaver is patient. He must never rush. You make a groove, patiently, and so it must not be rounded, but V-shaped. You don't need a hammer, you don't need a chisel."

"Just pressure," murmured Bernstein, taking another candy. He started to unwrap it, and then changed his mind. He tightened the wrapper neatly back into place and dropped the candy in his pocket.

"Knife must be exactly so," said Green, pressing the palms of his hands together with the tips of the fingers slightly opened, representing the groove where the knife would go. "And with a small knock"—he pulled his hands apart—"it cleaves." He folded the stone back into its paper packet and fished a leather pouch from inside his trousers. The pouch was fastened to his belt with loops. He inserted the packet and tucked the pouch back inside his pants. Then he replaced the skullcap with the fedora, exchanging them in the same fluid movement as before, shook hands all around, and walked out with the 82-carat cape yellow.

Green left Goldberg's building, crossed Fifth Avenue, and entered a lobby crowded with Hasidic Jews. In his tiny workroom on the twelfth floor, only the big Chatwood Milner safe and the motion sensors looked contemporary. The tools were the same tools cleavers have used for generations. Slabs of hard, brown cleaver's cement lay on the bench, ready to be melted in the flame of a spirit lamp and shaped to hold the diamond in place at the end of a cleaving stick. Green worked quickly, and soon the yellow rough sat tightly gripped in cement. Only the protrusion marked in ink stuck out. When he had studied the stone, and was ready to begin, Green would position the diamond over his cleaving box and, scratch by scratch, rubbing a sharp-edged cutting diamond against the larger stone, commence to make a groove. To the suggestion that he might fear the moment when he had to place the knife in the groove and strike, Green reacted forcefully.

"Never! The surgeon who is making the cut, is he nervous?"

He lifted his index finger from the stick. "Not." He put the stick down and pulled a tin box from the safe. Inside were scores of diamond packets. He riffled through them quickly, found what he wanted, and tugged it out. He flicked the paper open with his thumbs. A pink diamond nestled inside, the color so delicate it might have been blown into the crystal by a puff of breeze. Beside it lay a tiny sliver that Green had cleaved away. Green gazed at his handiwork. "I'm enjoying this very much," he said. Finally he folded the packet and wedged it gently back into the tin box among the other envelopes, and returned it to the safe. He added quietly, "I am very, very good."

The diamond trade is a struggle between rapture and calculation. Diamantaires like Goldberg try to see the jewel hiding in the rough. They gamble money that they can retrieve the jewel, and that it will be as beautiful when they get it out as it is inside their heads. When he bargains for rough, the polisher is betting on his ability to outguess the diamond.

Goldberg paid $60,000 for the stone he sent to Green. The stone was yellow, but how yellow? When they bought it they thought it was an M color, the last classification that could still be called "faint yellow." If in the end they obtained a 22-carat gem, graded M, they could sell it for $6,000 a carat, or $132,000. But if, when polished, the stone turned out to be not M, but N, it would descend into "very light yellow," less valuable than "faint yellow." The stone would lose $1,000 a carat in a blink, costing Goldberg $22,000. It wouldn't be the first time Goldberg had been disappointed by a stone.

Twelve years earlier a dealer had walked into Goldberg's office with what looked like a very good top-color white, and Goldberg paid him $250,000 on the spot. "It looked like a nice stone. I recall being quite excited by it. Maybe it had a gletz or some pique, I don't remember exactly. What I do remember is that it was a horrible surprise." Goldberg frowned as he cast back for the details of the stone. "It turned out to be almost uncuttable. It had knots all through it. I hadn't seen them. It was full of them. We had that stone on the wheel for six months. It took that long to grind it down. I made sixty cents on the dollar." He lifted one large hand and let it fall. "That's the game. You're in it or you're not. I know how to take a loss because I know how to take a profit."

While the 82-carat cape was still across the street with Green, another yellow diamond, a 50-carat fancy intense yellow, appeared in Goldberg's office alongside a 38-carat white. It seemed odd to Goldberg, because that summer it was hard to get large rough, and here were two good pieces. There was one catch. "Mr. Goldberg," the seller said firmly, "you can't buy one, you have to buy both."

Goldberg called in Barry Berg. They handed the rough back and forth. "There's a spot in that yellow," murmured Berg. They put the value of the parcel at between $525,000 and $550,000. Berg thought the seller might have difficulty placing both stones as a single package. He thought they should offer low, and suggested $340,000. Goldberg thought about it for a minute. If the offer was too low the seller might take offense, storm out, and not come back. Goldberg wanted the yellow. He thought it might be a smaller version of the Pumpkin, the 44.74-carat fancy vivid yellow that he'd sold to Harry Winston, and he didn't want it walking out the door. "Offer him three hundred seventy-five thousand," Goldberg said, "but tell him he can have the check right now. We'll put the money in his hand today." The seller left to think about it. When Goldberg didn't hear from him for a couple of hours, he thought it was a good sign. "If they're offended they call you names right away and tell you you're crazy, and they don't come back."

Three hours later the seller's broker called. If Goldberg would agree to pay the broker's 1 percent commission over and above the sale price of $375,000, the deal was done. Goldberg agreed, and the stones changed hands. Goldberg and Berg took a closer look at the yellow. This time they saw a cloud. They put the diamond back in its packet and sent it across Fifth Avenue to the cutter Ned Salzman. The way they had modeled the stone (decided on the shape and weight of the jewel they would try to get from the rough), the sawyer would have to saw through the cloud. Salzman took a look and called them back.

"Bill," he said to Goldberg, "this is a very explosive stone. You have to prepare yourself. I tell you honestly, it's a bomb, and it scares the shit out of me. That cloud in the middle—I don't know what's going to happen when the saw gets to it."

Goldberg plainly relished the situation, as if an old, beloved opponent had made a formidable move in the game of duplicity. Berg was not as sanguine, noting bluntly that there was a risk that the yellow would blow up on the wheel. Salzman did not attempt to saw it right away. He put it in his safe on a Thursday and it stayed there through the weekend.

On Monday Salzman came in and removed the yellow from the safe. He held a ten-power loupe to his eye and rolled the stone between his thumb and forefinger. He studied the faint blur inside the stone, a cluster of tiny bubbles. The sawyer put the loupe aside and placed the yellow in a polariscope, a system of filters arranged to detect idiosyncrasies of refraction. If there were something wrong, it would show up as blocks of red and green light. The more fractured the reds and greens, the more hazardous the stone. "This one was a bunch of tiny speckles," Salzman said, "very scattered, like grains of colored light. It was very dangerous. And that worst place, where the cloud was, that was where I had to saw."

Diamonds dictate the terms of their cutting. It is a commonplace of the trade that a cutter must think like a beam of light. He must imagine himself inside the finished diamond, and understand where the light should enter, and how it will bounce around inside. Few diamonds are flawless. Cutting helps conceal flaws. For example, cutters try to polish a stone so as to place the flaws in the upper part of the polished diamond, the "crown." There are fewer facets in the crown, which thus reflects light less than the lower part, or "pavilion." A flaw that is placed in the pavilion is reflected many times, and appears to the eye far worse than it really is. If the flaw is in the crown, an observer may miss it entirely.

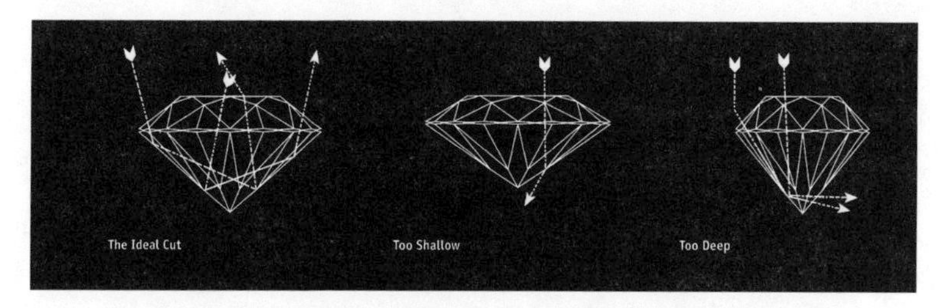

An illustration of the importance of proportion. In the well-cut diamond on the left, the angles are such that light entering the top bounces around inside the stone and comes back out the top, creating brilliance. In the shallow-cut diamond in the center and the deep-cut diamond on the right, the angles cause the light to leak from the bottom, reducing brilliance. (Dia Met Minerals)

"You have to know the stone," said Salzman. "I'm doing it fifty years. You have to know what that stone will give you. You saw for the finished stone. If you've got an off-color stone, you don't want to saw it in such a way that they have to polish it into a fancy shape, like a pear. It's hard to sell. It's kind of dead, with that color. So you give them something that can go to a round shape. A round can have lots of facets, more sparkle, and the [poor] color doesn't hurt so much."

Even with flawless, top-color white diamonds, the cut is all-important. If a stone is cut for maximum brilliance, it is said to be the best "make." Its proportions and angles will be such that it returns more sparkle to the eye than a stone of lesser make. Sometimes, for example, a stone is cut to preserve weight rather than to produce brilliance. Instead of a .75 carat stone of the best make, the cutter decides to retrieve a full carat with less glitter. Such decisions are reflected in the price. The cutter who cuts for weight throws out less rough than the cutter who pursues brilliance above all else. The buyer must pay for what the cutter discards. A top-color flawless 1-carat stone at Tiffany's costs $28,000. A few blocks down Fifth Avenue, the same color, clarity, and weight might sell for $20,000. On West Forty-seventh Street, a buyer might find a diamond with the same basic characteristics for $15,000. The difference in price relates not just to the jeweler's position in the retail geography of Manhattan, where Tiffany's occupies a fashionable block, but to the amount of rough the cutter was prepared to grind away to achieve his finished gem.

A sawyer like Salzman, then, must understand the nature of such deliberations. Salzman made his calculations about the yellow, marked the stone by drawing a line of black ink around it where he thought it should be sawed, and sent it back to Goldberg. Berg and Goldberg studied Salzman's mark, returned the stone, and told him to go ahead. He fixed the diamond in a clasping device called a dop, which holds the stone in place above the blade. Salzman turned on the saw, a vertical disc that spins at fifteen thousand revolutions per minute. Since only diamond is hard enough to cut diamond, the sawyer dressed the copper blade with diamond powder, mixing the powder in an oil solution and applying it with a steel roller to the spinning blade. Last, he adjusted a weight that would keep the diamond pressed to the saw, and began the cut.

The saw would excavate a slot 2/1000ths of an inch wide. At the atomic level, a canyon was being chewed through the crystal. A diamond crystal forms over millions of years, and its structure is annealed, or toughened, by exposure to heat that diminishes uniformly over more millions of years. But conditions do not always produce a perfect crystal, and the sudden intervention of a saw blade tearing through a diamond and generating heat can abruptly change the way stress is distributed in the crystal. If the crystal has internal weaknesses, it may rupture. The diamond will shatter into bits, or into a lacework of tiny cracks. When the cutter Gabi Tolkowsky was polishing the 273.85-carat Centenary diamond, he and the owners, De Beers, were so fearful of excessive heat that they rigged the stone with a special cooling system that kept the diamond's temperature below ninety degrees centigrade. But Salzman had no such cooling system; he was on his own. "It was a risky stone"—he shrugged—"so I worked a little slower. I didn't put as much weight on it, only a very light pressure, and I used less powder on the blade. You keep your fingers crossed. I'm not God."

Every fifteen minutes Salzman checked the yellow. He was

using a thinner paste on the blade, to reduce heat. With a thinner paste, he had to dress the blade more often. If he did not, the diamond would wear away the blade, or the blade would become too hot. The stone ran for seven hours straight. At 4:30 Monday afternoon Salzman shut down the saw, locked the stone away, and went home.

Next morning he began as before, mixing a special, lighter paste and balancing the armature that held the diamond so that only the barest pressure held the stone against the wheel. Hour after hour the hairline slot crept closer to the cloud of bubbles. Berg called several times. By Wednesday afternoon the saw had entered the cloud, and was chewing its way through. Salzman paced back and forth along the bench of whining saws, stopping at the yellow, watching for what he feared. Finally it happened.

"A gletz came in when he was in that bad area," Barry Berg said later. "He was halfway through the stone. He called and told us." Berg is tall and trim, and often wears a distant smile. "It's like baseball," he said of the yellow. "Sometimes you just get a bad pitch."

Salzman finished sawing the yellow on Friday afternoon. He sent the two pieces back to Goldberg. On Monday, Motti Bernstein started polishing the larger piece. It went well and yielded an 18-carat radiant-cut fancy intense yellow worth $12,000 a carat, or $216,000—exactly what they'd hoped for. Then they turned to the smaller piece. It retained much of the nest of flaws that had clouded the original stone. Bernstein gingerly put in the first facets. The gletz caused by the saw sat at the edge of the dangerous cluster of tiny bubbles. Berg and Bernstein decided that the risky section would have to come off, and Bernstein polished it away. When they'd bought the stone, Berg had hoped for a clarity grading of VS1 (very slightly included), which describes a stone with minor inclusions that are difficult to see. The appearance of the gletz dropped that rating to I1 (included), for a stone with an obvious inclusion, visible to the naked eye. They'd also had to cut the size, causing a loss of color intensity, dropping the stone from a "fancy intense yellow" to a "fancy yellow." Grade by grade, the value of the smaller piece of yellow fell. Where Berg had anticipated a 14-carat fancy intense yellow VS1, he ended up with a 10-carat fancy yellow I1. Instead of a jewel worth $140,000, he had one worth $35,000. The cloud had cost them $105,000. Moreover, the 38-carat white had polished into a 13-carat oval worth $130,000, where they had hoped for a finished stone worth $160,000. "Quite frankly," said Berg, "I'll be happy to get out of this without losing money. It happens. We take risks. Maybe we're not as careful as somebody else, but we have fun." The 82-carat cape yellow that Goldberg had bought on a whim for $60,000 took more than a month to polish, and finished as a 26.81-carat radiant cut that sold for $78,000.

◆

One of the most treacherous large diamonds of the twentieth century came out of the Premier mine on July 17, 1986, weighing almost 600 carats. De Beers put it away and swore everyone to secrecy. A few months later, in the autumn of the year, Gabi Tolkowsky, the Antwerp cutter, got a call from De Beers in London asking him if he was coming over soon. Tolkowsky thought this odd, because he visited London every month, as De Beers knew, for he came as their consultant. He said yes, he would be there, and the caller said no more.

When Tolkowsky arrived, an officer of De Beers met him at the door and took him to a large room on the sorting floor at 17 Charterhouse Street. Unusually, the room had been cleared of diamond sorters. Only one man accompanied Tolkowsky inside, and he carefully locked the door behind them. On a table was a single metal sorter's box. Tolkowsky crossed the room and opened the box, and there lay the large diamond. "It was so colorless," Tolkowsky recalled. "It was like water, but it was a diamond. And the whole appearance—I didn't dare touch it. The side I was looking at, it was looking like a brain. It was rounded. When I took it in my hand I was scared to crush it." Tolkowsky had good reason to feel that way, because the surface of the diamond was riven with cracks that ran down into the stone. As the craftsman could see at a glance, those cracks would be the cutter's ordeal.

The diamond returned to South Africa and no more was heard about it until March 11, 1988, when four hundred guests arrived in Kimberley to celebrate the hundredth anniversary of De Beers. The glittering company included prime ministers and presidents. Julian Ogilvie Thompson, then the De Beers chairman, made a welcoming speech, which he ended with the surprise announcement that De Beers had found "a diamond of 599 carats which is perfect in color—indeed, it is one of the largest top-color diamonds ever found. Naturally it will be called the Centenary Diamond." De Beers asked Tolkowsky to cut the Centenary, and he agreed, and so began the most remarkable tactical assault on a diamond ever recorded.

In late 1988 Tolkowsky and his wife, Lydia, closed their house in Antwerp, left their two black Labrador retrievers with their children, and moved to Johannesburg. De Beers assembled a team of polishers, engineers, and electricians to assist the cutter in his undertaking. An underground room at the Diamond Research Laboratory was set aside and wrapped in the tightest security, including its own guards. The room itself was engineered so that no vibration would imperil the delicate operation of the cutting. Thermostats were set to maintain a constant temperature of about sixty degrees—pleasant for the diamond but cool for the men.

Having made so public an announcement of the stone's size and quality, De Beers had put its own prestige on the line. If any accident befell the diamond during polishing, the shame would be as much De Beers's as Tolkowsky's, and the company spared nothing in shielding the cutter. When a rush of reporters came looking for Tolkowsky, De Beers flew him and his wife to a villa in Cape Town, where the couple remained until the hue and cry died down. Later, when De Beers brought them back, the security staff remained particularly vigilant, fearing a kidnap attempt on Tolkowsky's wife with the Centenary demanded as ransom.

Tolkowsky began by making an exhaustive study of the diamond. Every day he went to the underground chamber and examined the stone. The walls of the bunker were painted a pale green, like an operating room, so that the eyes of the surgeon, Tolkowsky, would not be bothered by glare. He noted every crack and blemish of the stone, and thought long and deep about it. When he recalled this, Tolkowsky seemed to be describing a person. "There were some inclusions coming from the surface of the stone into the stone, pressure cracks, which showed that this diamond, when it created itself, had resisted huge powers which tried to crush it, and this fantastic crystal had resisted the penetration of the powers, and stopped them."

Tolkowsky came from a long line of diamond cutters. His great-grandfather, Maurice Tolkowsky, emigrated from Russia in 1880, settling in Antwerp and starting a diamond factory. Several of Maurice's brothers were diamond cutters too, including Isidore, whose son, Marcel Tolkowsky, a mathematician and diamond cutter, invented the modern brilliant cut in 1914 when he published a theoretical treatise that laid down the proper proportions for maximum light return. The fifty-eight facets and precise angles established by Marcel Tolkowsky have remained the basic system for producing brilliance in a diamond, and the subsequent, more-faceted cuts that have appeared are but elaborations of Marcel Tolkowsky's classic cut. Gabi Tolkowsky, Marcel's great-nephew, has himself created diamond cuts, including a series of flower cuts developed for De Beers.

Gabi Tolkowsky studied the Centenary diamond for a year, discovering the magnitude of the challenge. As he scrutinized the larger cracks with a microscope, he saw, at the deepest point of penetration, networks of much tinier cracks and, at the edge of each of these tiny cracks, a bubble. It was these infinitesimal bubbles that frightened Tolkowsky most, for a British researcher, opening just such a minute flaw in a diamond, had detected moisture. As Tolkowsky thought about this, a phantom rose in his mind. "What if one of them, one of those micro bubbles, contained water, and I was cutting the stone and creating a temperature of four hundred degrees in a city where water boils at

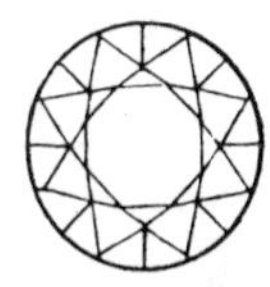 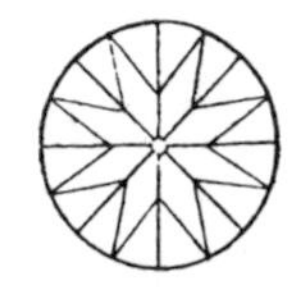 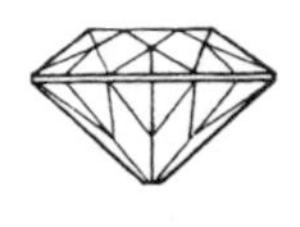

Left to right: The top, bottom, and side views of a diamond polished into the classic brilliant cut devised by Marcel Tolkowsky in 1914.

ninety-three point seven degrees—what would happen to this small tiny bubble? Perhaps it would open the door to what nature had not succeeded in doing!"

With De Beers engineers, Tolkowsky developed a cooling system for the diamond. The rough gem would be clamped into a copper cup. The cup had an inner and an outer skin, with water passed between them at a constant temperature of 60.8 degrees. Diamonds conduct heat well, and heat generated by the friction of cutting would be passed into the copper and, in effect, carried away from the diamond by the flowing water. But the system was untested, and Tolkowsky would not entrust the Centenary's survival to experimental technology. By a stroke of luck, another large diamond had recently been discovered in the Premier mine. It weighed 755.5 carats, but its brown color made it far less valuable than the top-white Centenary. The brown too was a dangerous stone, riddled with cracks, and therefore an excellent test subject. Tolkowsky started polishing it in May 1988, and within a month had ground away almost 60 carats. The diamond, christened the Unnamed Brown, held up perfectly. With no more reason to delay, Tolkowsky fixed the Centenary into its special apparatus and began to work.

Although advanced technology supported him, Tolkowsky approached the great diamond with the oldest tools of his craft. He cemented a smaller, sharp-edged diamond to the end of a cleaving stick and began the laborious, time-honored process of kerfing. A kerf is a notch rubbed into a diamond with another diamond, usually preparatory to cleaving. Tolkowsky did not propose to cleave the rough, but to rub away the cracks, one at a time, trusting the Centenary only to the touch of his own hands. He worked for 154 days. In removing the cracks, he took off 80 carats, and transformed the diamond into a roughly rounded shape about the size of a hen's egg. It weighed 520 carats, and was ready to polish.

When Tolkowsky had learned he was to cut the Centenary, he had asked De Beers to cast fifty models of the rough stone in plastic resin. When he had them, he garbed himself in an apron and goggles, fastened a mask over his nose and mouth, and began to experiment by polishing the models. A cloud of plastic filled the air as he ran off shape after shape on the wheel. He made almost forty shapes, some conventional and some not. Among them were a series of hearts, and for a time Tolkowsky believed that a heart would be the diamond's final shape. But something kept nagging at him. He began to think that a conventional shape would not do. He found himself wanting to challenge the diamond. He decided to aim for a 300-carat polished. In this endeavor he would have to make his way alone, unsupported by the mathematical comforts of his great-uncle's formula. The diamond taking shape in his imagination was not mapped anywhere. It was unique.

When Tolkowsky had finished making models, De Beers decided to show them to Margaret Thatcher, the British prime minister, who was about to visit the London office. They set aside a private room and laid out a selection of Tolkowsky's plastic "diamonds" on a velvet cloth. There were pears and marquises and ovals and hearts, and a few of the large, swelling forms, like plumped-up cushions, that the cutter had evolved for the Centenary alone. "She had gray eyes," Tolkowsky recalled of Thatcher, "metal eyes, and she looked at all the models, and I explained what I am doing, and she said, 'If I were you, I'd make this one.' " The model Thatcher touched with her finger that day in London was very close to the shape Tolkowsky settled on, and when he told the story to a De Beers director, the executive said: "Well, Gabi, it seems more and more we have to ask the women what to do."

In March 1990 they put the Centenary to the wheel.

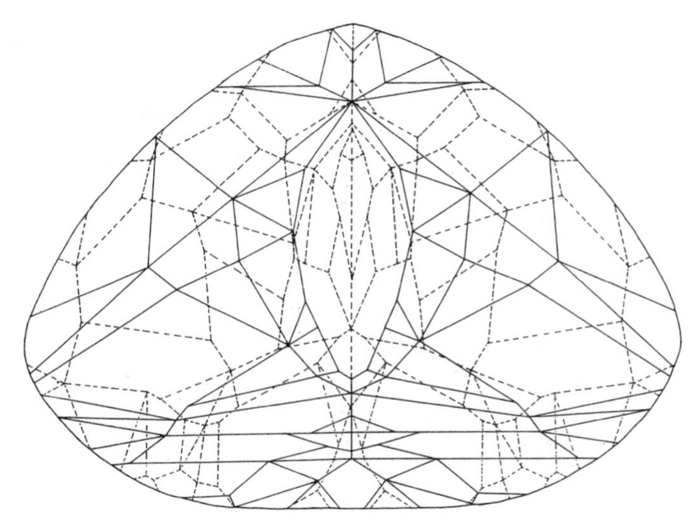

Facet plan of the Centenary diamond showing the complex scheme that yielded 247 facets. (De Beers)

Tolkowsky and his assistants ground their way into uncharted territory, toward a shape no other cutters had essayed—an enormous mound of facets. They lived with dread. Although Tolkowsky had removed the major cracks, the tinier cracks remained, and with them, the microscopic bubbles. The cutters did not know what, if anything, the bubbles contained, but had to venture on into the stone, hour after endless hour, grinding away the diamond that concealed the diamond.

Fans drew off the diamond dust and expelled it from the chamber. A pair of sensors on the diamond monitored the temperature. As soon as the heat rose to eighty degrees, they stopped and waited for the coolant to reduce the temperature before they began again. The work consumed every member of the team. Social life contracted. They saw less of their families. The diamond changed the habits of lifetimes. Men who liked parties gave them up, keeping their heads clear for the diamond. They gave up sports that posed any risk of injury, so as not to damage fingers or eyes. The stone absorbed their energies, as if the cutters' own strength were being ground away on the wheel. Tolkowsky said that polishing the diamond tortured him. He dreamed about the stone. "I looked at the diamond all night long and the diamond looked at me."

They polished it for a year. Every day began with the same exhaustive ritual. At eight o'clock, the cutting team gathered to conduct a painstaking check of all their tools. They verified the polishing angle. They took the temperature of the diamond coolant. Fixing the diamond in the holding-and-cooling device took about an hour, and sometimes as long as three hours, depending on the facet to be polished.

Facet by facet the jewel emerged. De Beers executives noted the progress, and as March drew round again they began to agitate for an end to the long trial. It had been three years since they had announced the discovery to the world, and the company wanted a polished jewel to show for it. A director came to Tolkowsky and said: "Now, Gabi, the game is finished. We have decided the stone must be done before the first of May. It must be in London before the first of May."

"You can say what you like," replied Tolkowsky, "but the stone must agree."

The director shook his head. "It's three years and that's enough. *Before* the first of May."

The diamond Tolkowsky finished that April in Johannesburg was a masterpiece, the magnum opus of a virtuoso cutter. You might even say it was more than that, and proceeded not just from the heart and mind of a single man but from the generations of a great Jewish diamond family. The jewel swells up in a perfect heap of facets, impossibly intricate and shimmering. A gale of light thrashes around it. There are seventy-five facets on the top and eighty-nine on the bottom, and the girdle is a breathtaking pattern of eighty-three facets. Said Nicky Oppenheimer when De Beers unveiled the diamond: "Who can put a price on such a stone?" They insured it for $100 million.

As for the Unnamed Brown that Tolkowsky had affectionately called his "ugly duckling," like the bird in the fairy tale, it turned out to be a swan. When Tolkowsky had finished using the Unnamed Brown as a stand-in for the Centenary, he made a plan to polish the colored diamond. The master cutter Dawie du

Left: Gabi Tolkowsky examining the Centenary.
Right: A close-up of the Centenary. (De Beers)

Plessis was given the assignment. He took a year to polish the gem into a cushion shape. In the rough, it had shown a mysterious glimmer at its heart, and this promise now shone forth in a blaze of golden light. At a finished weight of 545.67 carats, the Unnamed Brown was bigger than the Great Star of Africa. Tolkowsky's ugly duckling had become the largest polished diamond in the world. It could hardly remain "unnamed," and De Beers rechristened it the Golden Jubilee. A group of Oriental buyers purchased the diamond for an undisclosed sum, and in May of 2000 presented it to the king of Thailand's daughter, Princess Maha, who received it for her father.

◆

Great diamonds rain their glitter onto the whole trade. The bravura performances of cutters like Tolkowsky and William Goldberg become the stuff of diamond legend. When Goldberg bought the Queen of Holland diamond from Cartier in London the 136.25-carat cushion-cut gem was one of the most perfect white diamonds in the world, with that tint of blue that distinguishes the rarest whites. Its provenance was faultless: Cartier had bought it from the heirs of the Maharaja of Nawanagar. Few diamantaires would have dared to tinker with such a diamond, but Goldberg thought he could improve it and he sent it to the wheel. They shaved off a third of a carat to give it a bit more sparkle, and sold it for $7 million.

Goldberg and Tolkowsky pratice their sorcery at the top end of the trade. Ironically, though, the most important advances in poloshing have taken place at a much humbler level, among goods so small and dim that they look like heaps of sand. In New York there are some 250 cutters, in Antwerp about 1,500, and in Tel Aviv perhaps 3,000. But in Bombay, hundreds of thousands of cutters hunch at diamond wheels, and their exertions have revolutionized the definition of a jewel, so that many diamonds that were sold into the industrial market are now polished into gems, increasing the financial viability of mines from Botswana to the Barrens. In the course of this great change, tremendous resources of power and money have moved into new hands, and the control of a staggering share of the world's diamond business has moved back to the place where it all began—India.

11

Rosy Blue

At eight o'clock on a hot morning in February 2000, a group of men waited in front of the Bombay domestic air terminal, in a space reserved for the arrival of important travelers. As they stood chatting among themselves, a chauffeured sedan turned in and stopped beside them. The thirty-eight-year-old man in a sports jacket and open-necked white shirt who climbed out of the backseat was Russell Mehta, whose family runs Rosy Blue, the largest diamond-polishing company in the world. Taken together, the dozen friends at the airport that day represented a polishing powerhouse that sends to the wheel some 680 million diamonds a year.

The world's diamond polishers sold $11 billion worth of finished goods in 2000. India had 50 percent of the business by value, 80 percent by volume. This large discrepancy between value and volume is due to the nature of the goods polished in India. Of the hundreds of millions of diamonds a year that pass through their factories, most are very small. The Indians polish them in great volume by employing hundreds of thousand of cutters in Bombay and Surat. Some people in the diamond business sneered at the Indians when they arrived in the trade, but they do not sneer now. In only a few decades, Indian manufacturers have reduced the older cutting centers of Antwerp and Tel Aviv to shells of what they were. The friends gathered at the Bombay terminal were the visible symbols of that success, of families who had transformed themselves from cottage-industry practitioners to multinationals in a generation.

On that February morning the group was flying to Surat, a city in the state of Gujarat, to represent their families at a wedding. Indian diamond traders famously hate to take time off from work. Seven-day weeks are not uncommon in the diamond quarter in Bombay. But this wedding was as much a business as a social obligation, when one of their own number would present a spectacle by which others could judge his success, and by attending the event, the Bombay millionaires were celebrating not only their compatriot's attainment but the ascendancy and pride of the Indian diamond trade.

The men made their way into the terminal, cleared a rudimentary security check for VIPs, and exited the building into waiting vans. A jeep with a flashing light led the little cavalcade out onto the runway, then turned off toward a group of private hangars. Two twin-engine planes waited on the apron, and Mehta and his friends filed out of the vans and climbed aboard. A few minutes later the planes taxied out, ran down the runway, and lifted into the humid air. Several miles of dun-colored hovels drifted by below until the aircraft reached the glittering waters of the Arabian Sea and turned north. After an hour's flight, the aircraft sank down onto the plain of Gujarat and landed at a short strip, the airport of Surat.

Half a dozen white cars waited on the apron. A large man with a barrel chest and a flowing mustache and the cuffs of his spotless white shirt buttoned at his wrists hurried up to Mehta, pressed his hands together in greeting, and led him to one of the cars. He wedged his own massive bulk into the front seat and the car went bouncing down a dirt road and onto a four-lane highway, and sped toward the city.

Mehta took out his cell phone and held a short conference in rapid Gujarati in which the word *Antwerp* occurred twice. The car turned off before the city, made its way through a suburb of white buildings, and turned through concrete gates into the gardens of the Holiday Inn. The hotel occupied a verdant enclosure facing the Tapi River, which flows out of the Gujarat plain on its way to the sea, ten miles away. From the terrace of the hotel, the view was of a Hindu temple across the river. Herds of water buffalo wound slowly up the shore. Narrow fishing boats with scraps of sail fled downriver on the ebbing tide. The scene was idyllic, like a postcard of old India. Nothing suggested the nearness of an industrial city of 3 million people, where thousands upon thousands of diamond polishers crouched at their wheels in pursuit of a trade whose profits, at this moment, were helping to fill the Holiday Inn with wedding guests.

People had come to the wedding from Antwerp and Tel Aviv as well as from Bombay. They paid for nothing while in Surat, and were ferried everywhere in private cars supplied by Arvind Shah, the father of the groom. By the time Mehta and his friends arrived the party had been going two full days, with music and dancers and feasts. Now the third and final day of the wedding celebrations had arrived—the day of the marriage itself.

From the Holiday Inn the guests proceeded to Laxmi Villa, Shah's mansion. A throng jammed the courtyard. Arvind Shah, smiling and harried, struggled through the crowd from guest to guest, clasping their hands in his own. In a room choked with women sat Chirag Shah, the groom, with girls in saris giggling around him and decorating his face with pigment. Mehta and his friends filed off in the opposite direction, to a terrace shaded by awnings. Food and drink were spread on a long table. One by one, the men sat in a chair while an attendant wound a crisp white turban tightly into place. Every minute a cell phone would ring, or two or three at once, and a turbaned guest would detach himself from the press, peel up the edge of his turban, and press the phone to his ear. Soon trumpets blared from the street and drums were pounded. The guests streamed out into a hot, packed lane, and the wedding procession began.

The narrow thoroughfare bulged with revelers. The bandsmen wore plumed helmets and uniforms of blue and gold. The music crashed and blared in the little street with a deafening force. Arvind Shah threw himself into a fevered dance. His turban flew off and his dark hair shook in the sunlight. Guards with red sashes and scarlet cockades in their berets strode along the edge of the procession, marshaling traffic aside with their batons.

The party moved a few feet, then paused to dance. Hands reached out to drag guests at random from the throng. A banker from Antwerp flailed about in a dark blue suit, his turban slipping perilously. A diamond polisher from Tel Aviv shouted a final word into his cell phone, just managing to snap it shut before his host yanked him into the dance. Arvind Shah himself, a fury of jubilation, had a cell phone in his hand as he capered along. So they went like that, up the dusty road, connected to London and New York, until they came to the marriage ground.

The party surged into an enormous tent. Huge ceiling fans hummed from crossbeams. Hundreds of chairs faced a dais. Booths with cold drinks and mango ice cream lined one long wall of the tent. At the rear of the cavernous space, white tables stood in a VIP enclosure marked by a wooden fence. Near the dais, a quartet of women sat on a low table and sang. The groom was led to a throne beneath a canopy, where he sat and waited while the bride, a trapeze of jewels on her face, was led toward him. In the enclosure at the back, stewards in white jackets served food from silver chafing dishes, and Mehta and his associates talked among themselves.

◆

The origins of the trade that has provided wealth to the Mehta family, to Arvind Shah, and to most of the wedding guests may be found in the old diamond quarter of Surat. Tiny polishing factories are secreted here and there in the warren of streets. An alley leads into a large courtyard packed with bicycles. A gallery runs around the courtyard on the second floor, and guards with old bolt-action rifles lean against the railing. The buildings are packed with polishing factories, sometimes no more than a single wheel, with four members of a family crouched around it in the half-light. The walls are concrete and the floor is concrete; they squat on the bare floor and polish the cheapest, smallest diamonds in the world.

The goods are usually brown and may weigh as little as half a point. One hundred points equals one carat, so a half-point diamond weighs one two-hundredths (0.005) of a carat. An adult needs eyeglasses even to see it, never mind to polish it. Children may polish such goods because their eyes are keener. These least of all diamonds are polished with only a few facets, something to catch a bit of light and flash it back. They are used to brighten the surfaces of cheap jewelry. Whiter, better goods are polished into a full brilliant cut, with fifty-eight facets and a finished weight of one point, or one one-hundredth (0.01) of a carat. On this work the cutter makes a commission counted in pennies per stone, which he increases by switching the goods.

The switching market of Surat operates openly in a pair of intersecting streets in the heart of the old town. One-room weighing-houses equipped with scales line the streets. The narrow thoroughfares are jammed from side to side with diamond traders. They cluster about each other's goods, comparing and arguing and bargaining. Finally the deal is done, and the traders resort to the scales, and everything is added up and made complete, and the cutter goes away with exactly the weight of goods he came with, except they are a little worse, and the difference sits in his pocket, in cash, while the cheaper goods go back to his client.

The larger dealers and polishers who farm out goods to the

Diamond trading in Surat. (Matthew Hart)

cottage polishers accept this slight loss as a cost of doing business. The profit margins for cheap goods are much higher than with many better goods, and can accommodate the losses entailed in switching. The cheapness of the cheapest rough means that it can be cut quickly and even badly. A poor yield does not much affect the markup, which is in pennies anyway, and depends on volume for its attraction. The small cutter, then, goes home with extra profit. In time he may save enough to buy another wheel and hire someone else to squat on the floor and polish the goods.

The Indians invented this end of the diamond trade, the lowest end. Before the establishment of the India trade, an average of 80 percent of a mine's production was industrial diamonds, with only 20 percent as gems. Nowadays, a more common split would be 40 percent industrial and 60 percent gem. Even that division is misleading, because industrials may include the so-called "near gems," and Indians polish those too. The industrial market is supplied by synthetic diamonds anyway, so that the Indians have transformed the waste of the diamond-mining world into jewels. It is the astonishing multitude of polishers who accomplished this transformation.

Some four thousand factories polish diamonds in Surat. At the wheels there and in Bombay sit five hundred thousand diamond cutters, or perhaps seven hundred thousand. The size of the labor force fluctuates according to demand for polished and supply of rough. The Indians themselves do not know the exact population of artisans at work at any time. This business grew from nothing to its present state in forty years. At first, the Indians stuck to the cheap goods and avoided attacking the business of Antwerp and Tel Aviv. Now they cut whatever they can get their hands on, and as their ability to produce fine polished has advanced, so has the amount of better rough put into the sight boxes of the biggest Indian polishers.

In Surat today live many rich, ambitious diamond polishers. At the marble headquarters of Karp Diamonds, in the suburb of Athwa Lines, the sorters work in spotless, air-conditioned rooms. When Kishor Maldar, one of the proprietors, wishes to visit his distant factory, a helicopter lands on the roof to pick him up. The factory is 250 miles away, in northern Gujarat, in a compound with houses for the polishers, along with schools and stores. Parag Shah, Maldar's friend, runs an immaculate factory in Surat, where the movement of every stone is monitored and the yield watched by Parag Shah. Entrepreneurs like these have helped create a concentration of polishing power and commercial exuberance that has eroded the importance of the cutting centers of Antwerp and Tel Aviv.

◆

The rise of the Indian polishers has reversed the trend of centuries, in which India declined as a diamond power. Some thirty mines once operated in India, from about the middle of the fifteenth century until the nineteenth century. The largest digs were on the Kristna River in the state of Hyderabad, where, in the seventeenth century, Jean-Baptiste Tavernier reported sixty thousand men, women, and children at work. There are no production figures for these mines, but although they produced many of the world's greatest diamonds, such as the Koh-i-Noor, overall output is thought to have been extremely low, and as a diamond center India was eclipsed by Brazil and, later, South Africa. In the field of cutting, too, India was left far behind by European cutters, whose techniques were developed in response to a taste for brilliance that historically did not exist in India. India's diamond fortunes did not begin to turn until 1947, when the Indian government, newly independent and alarmed at the state of its foreign reserves, prohibited the import of certain luxuries, including polished diamonds. They made one crucial exception: those who had imported polished diamonds could now import rough, with the proviso that they then export polished.

But what goods could Indian cutters polish? The skills developed in antiquity had suited the old Indian taste of scraping off an edge here and there to allow light into the stone. There was no tradition of polishing for brilliance. A cottage industry of polishers survived in Surat, polishing semiprecious stones. The few Indian dealers in Antwerp at the time, whose business had been to export polished goods from Belgium to India, knew it would be mad to send high-value rough into a country whose polishers were certain to destroy its value. They hit upon the idea that was to revolutionize the trade. They would send home for polishing the cheapest rough they could buy, the so-called Antwerp rejections. Even if the Indian cutters mangled the goods and produced a low yield of polished from the rough, the price of the goods was so low and the labor so cheap that the owners would still make a profit.

Positioned to take advantage of this new trade was a struggling family of pearl dealers and diamond polishers in Bombay, the Mehta family, whose antecedents had been jewelers to the nawab of Palinpur, in Gujarat state. The Mehtas' pearl business was not thriving. Their only diamond business was the repair of old-cut diamonds. Ramniklal Mehta decided to enter the trade in small diamonds that had begun to grow so rapidly around him. He bought a parcel of rough on the street in Bombay, took it to a cutter to be polished, then sold the polished to an exporter. He made money, so he bought another parcel and did it again.

Then Ramniklal's brother-in-law, Bhanuchandra Bhansali, entered the diamond business too, with a single wheel. The conditions in which he worked resembled those of the cutters in Surat. A package of rough would arrive and a deal would be struck, and Bhansali would go to work in the dreadful light, sitting on the floor at the spinning scaife, putting the first of fifty-eight facets onto a diamond the size of a grain of sugar. When he was done, he would return the parcel to his client, collect his fee, receive more goods to polish, and go back to work. Bhansali prospered. He bought another wheel, and then three more, took a second room, hired more cutters. His business grew with the growing Indian trade. In 1958, Arun Mehta, Ramniklal Mehta's son and Bhansali's nephew, left college after only a year and entered the diamond business of a family friend.

It is fair to say that Arun Mehta had a head for diamonds. The young man learned to clean and sort the goods, and then to sort them again into those specific assortments most in demand by his employer's customers. To sort diamonds skillfully is to increase their value, because the sorter separates the diamonds into lots, each of which includes only those qualities and sizes that a specific client wants, and for which he will pay more.

Arun Mehta plainly loved to sort diamonds. He studied diamonds day and night. He bent over piles of rough and sorted through them stone by stone, examining the diamonds with his loupe. Then he would rake a pile of polished across the desk and sort that too. By this time his uncle's polishing business had grown to seven wheels. In his spare moments, Arun would find a wheel and practice polishing diamonds. Before long he was buying polished on his own, re-sorting it into marketable lots, and selling it into the expanding trade. Things went well until, one day in 1959, the Mehta family suffered a blow. "Our father had become a partner in a diamond company," says Dilip Mehta, Arun's younger brother. "He came home one day, he was very down. It was very silent. I didn't know what was happening. He was looking miserable. My mother asked what happened, and his partners had fired him. My brother Arun held his hand and said,

'Don't worry. It'll be okay. You won't regret it.' And I remember that to this day."

In the Mehta family they refer to Arun as "the rock." The year after his father was fired, Arun borrowed $200 from him, and with his uncle and another partner started B. Arunkumar & Co. The *B* stood for Bhansali. Arunkumar is the formal version of Arun. They established a small factory in Bombay. They bought rough, polished some on their own wheels, and farmed some out to cottage polishers in Bombay and Surat. The profits went straight back into the business. They bought larger quantities of rough. Arun's brother, Arshad, joined the firm, and his father came in too. No doubt the Mehta family would have plugged away, growing somewhat richer every year, but for a series of three remarkable opportunities, which they seized, and which vaulted them into the forefront of the trade.

The first break came in 1963, when Chinese troops made an incursion into India. In the ensuing panic, Indian diamond prices fell 20 percent in eight weeks. Arun Mehta realized there was no rational reason for the fall. The diamond price, quoted in dollars, remained stable everywhere else in the world. The crash was only local, and emotional. To Arun Mehta's mind, the goods were basically on sale. He bought up diamond stocks and resold them overseas. The family made windfall profits. When the Indian government saw that the infant trade could produce such impressive amounts of foreign exchange, it began to encourage the import of rough, and supplies increased.

The Mehtas (and other Indian polishers) were handed another opportunity three years later, in 1966, when the Indian central bank devalued the rupee by 57 percent. At a stroke, the Indians found their wages and overhead slashed by more than half compared to their competitors. This happened at a time of strong world demand for polished, which the older cutting centers of Antwerp and Tel Aviv could not meet by themselves. Indian goods, which had improved in quality, poured out to fill the gaps.

The third break for the Mehtas came in 1969. Indian dealers

were by then buying huge amounts of rough, and De Beers acknowledged the increasing importance of the India trade by inviting nine Indian companies into the London sights. One of these was B. Arunkumar & Co. The Mehtas had arrived.

With their expanded business and position, the Mehtas concluded they could not remain solely in India. The capital of the secondary trade in rough was the old Flemish river port of Antwerp. Dealers with sights resold portions of their goods in the city, and the old diamond quarter was the world's main free market, or non–De Beers, source of rough.

When Dilip Mehta arrived in Antwerp in 1973 to open the family's beachhead in the city, the diamond quarter was overwhelmingly Jewish. Mehta soon ran up against this fact in an unexpected way. The family had decided to mark its arrival with a new company name. They hit upon the Indian term for the most valuable shade of fancy blue diamond, a name that would sound exotic but not alien. Dilip Mehta made all the arrangements, prepared the documents, and went to an Antwerp notary to execute the incorporation papers. The notary asked for the name of the new enterprise. "Rosy Blue," said Mehta. The notary looked up in surprise. "Rosenbloom?" he said.

No one in Antwerp would have to be told today how to spell the name. Rosy Blue occupies the largest offices in the Antwerp diamond quarter, in the smartest building, on the top floor. The rooms are chaste and stylish, with pale slate floors and bleached wood walls and contemporary art. There is a feeling of modish gentility, which is misleading. At Rosy Blue the only object is to sell diamonds, an activity carried on at times with high emotion. Once, when shouting erupted outside the conference room where Dilip Mehta was talking to a visitor, Mehta smiled happily. "There's a lot of screaming going on," he observed contentedly. "They [Rosy Blue's traders] are very focused people." He waited for the shouting to die down before he added, "They work seven days a week."

Rosy Blue employs twenty-five thousand people in nine countries, trading in polished and rough, cleaning and sawing the

Polishing diamonds at a Rosy Blue factory. (Rosy Blue)

goods, polishing diamonds on the wheel. At any moment there might be another fifty thousand independent cutters contracted to finish the company's goods. Diamonds pulse through Rosy Blue like blood through the veins of a large and vigorous animal. Today the India trade looks inevitable, as if it could never have failed to prosper, but had merely to be set in motion for money to come pouring out. Arun Mehta confesses that he thought so himself.

"Until nineteen seventy-nine," he said, "I did not think it was possible to lose money in the diamond business. We made money selling the goods, and the government gave us licenses to import more rough. Because we earned foreign exchange, they also gave us quotas to import certain controlled items, like machinery, which we could sell for a good profit." But in 1979, rapid inflation in the United States attacked the value of the dollar, and the rupee gained against the dollar for the first time in history. The Indians owed money to their banks in rupees, but sold their diamonds to people who paid in dollars. Those dollars were now worth less. The effect of this was that the polishers owned diamonds for which they had paid more money than they could now get. No diamond could be sold except at a loss.

The crisis sailed into the India trade like a rogue monsoon. It was an exact reversal of the situation thirteen years before, when a devalued rupee had presented the Indians with a bonanza. This time the math was against them. Many Indian manufacturers began to hold back stocks from the market. They refused to sell. They laid off cutters and suspended operations. Fear spread through the trade. The overhang of debt, with no revenue to service it, grew larger. Arun Mehta summoned a family council. "I said to Dilip, the diamond industry is based in U.S. dollars. The price of rough is in U.S. dollars. That's not going to change. So why not take a onetime loss by selling our existing stock? We'll take the hit, then come back in."

The Mehtas sold off all their goods, took a heavy loss, and hurried back into the market, buying rough as fast as they could and rushing it to the wheel. With their competitors idled, Rosy Blue began to rake in profits. Some business rivals complained. In Antwerp, people said openly to Dilip Mehta's face that Rosy Blue was acting unfairly. As Mehta himself recalled, a rumor circulated that the company was falsifying its books and would go broke. Rosy Blue just bought more rough and polished it and sold it.

Another crisis hit the diamond trade at this time, when some sightholders in Antwerp, Tel Aviv, and New York began to speculate in rough. Instead of selling their London boxes straight on into the market, they held back important qualities and sizes, creating a demand that drove up prices. De Beers watched this with dismay and, one would suppose, anger. After all, the manipulation of the diamond price had long been a De Beers prerogative. That is why they *had* a cartel. Now here were mere dealers usurping the practice for themselves. It was not simply that De Beers feared the speculation would end in a price collapse, but that diamonds it was selling for $100 a carat might be sold by someone else for $200 a carat. In this scheme, De Beers would "lose" $100 a carat. Clearly the dogs had taken over the kennel.

De Beers acted swiftly, placing a 30 percent surcharge on rough. Faced with this sudden, unexpected levy, dealers had to dump the hoarded goods. The value of a 1-carat D-flawless, which speculators had driven up to $70,000, fell to $10,000. The price of the hoarded goods collapsed in a heap. Many companies failed after this episode, but not Rosy Blue. The India trade rested on small diamonds, in which there was not much speculation, and Indian diamond capital survived the punitive action intact. Unharmed, the Indians took advantage of the chastened market. The affair illustrated the central irony of the India trade: out of the cheapest diamonds had come great financial strength. This strength threatened the Indians' business rivals but enriched another diamond sector—mining.

Bombay's appetite for small goods meant that a mine's production now contained more gem-grade stones than it had before. Hundreds of millions of carats of industrial diamonds were changed into gems by the India trade. The garbage of the mines was transmuted into jewels. This new market changed the rules for calculating the viability of a diamond deposit. The Indian cutters, at their wheels in ill-lit rooms, transformed not only minuscule rough diamonds but a whole industry. The biggest diamond mine in the world, the Argyle mine in Australia, would not exist without the India trade, and thus would not have led De Beers into a miscalculated fight.

◆

In 1986, its first full year of operation, Argyle produced 29 million carats—40 percent by weight of that year's world production of natural diamonds. If Argyle's diamonds had been of average quality, its owners would have found themselves, at a stroke, the equals of De Beers. But except for its pink diamonds, the Australian mine's stones were small and brown. Where Jwaneng, for example, produced rough worth an average of $100 a carat, Argyle's average was a lowly $11. This was bad news for Argyle but good news for India, since no one else could polish such diamonds.

Argyle was a joint venture of Rio Tinto and Ashton Mining Ltd., an Australian company. Initially, the venture partners contracted to market their diamonds through the cartel. Early in the contract, however, according to reports in the Australian press, a dispute arose about De Beers's valuation procedures. As part of the arrangement betgween De Beers and Argyle, De Beers had twenty diamond valuators on station in Australia to value representative samples of the run of mine and extrapolate the price to be paid by De Beers for the whole production. As detailed in the press, Argyle disputed the values that De Beers was assigning to the goods, and an angry confrontation between buyer and seller arose.

Argyle would not comment publicly on the reported dispute. When the story was later put to De Beers, a spokesman for the company wrote in an e-mail: "As you will know, some valuation criteria are based on a subjective view of wher an inclusion lies in a stone, or how badly a craxk or gletz affects a rough diamond. To accommodate this there is an accepted difference between valuations of the same parcel, usually around 15 percent. This was the case early on during our contraact with Argyle; however, it was sorted out amicably."

Nevertheless, in 1996, ten years later, Argyle abandoned the cartel. De Beers began to sell huge amounts of low-end rough into India, stuffing the Indian sightholders' boxes to the extent that the Indians would have little money left to spend on Argyle goods. As much as $200 million in cheap rough went into a single sight, depressing the price of Argyle's goods by 25 percent.

De Beers attacked the confidence of the Indians' bankers too, by making dire predictions in the annual Bankers' Booklet written by De Beers staff and distributed in Bombay. "It is now felt," De Beers warned, "that Argyle's decision to market independently will inevitably cost the industry substantial inventory write-downs." To make matters worse, large volumes of Russian technicals were also flooding into India at this time, further choking the market with goods whose price kept falling.

De Beers's ire may have been partly fed by events in Canada, since by 1996, the year that Argyle broke away, it was becoming clearer that the cartel would not corner Canada's rough. Perhaps De Beers perceived a need to publicly discipline Argyle, and show

how tough a game Johannesburg could play. The plummeting prices for Argyle's rough forced the mine perilously close to the break-even point. De Beers's actions shook the Bombay polishers, and their confidence in the future declined.

Argyle knew the battering would not last. It mounted a campaign of its own in Bombay, calmly insisting to the polishers and bankers that the business fundamentals were sound, and providing projections that indicated a recovering demand. This reassured the Indians, who had already made large investments adapting their wheels to the troublesome, hard-to-polish Argyle goods. If De Beers had made a mistake, it lay in underestimating the India trade. The torrent of cheap rough did not drown it, but made it swell. The long-term effect of the punitive action was to create yet more demand for Argyle's diamonds. De Beers seemed to acknowledge as much four years later, in 2000, when it made a surprise offer to buy out the 40 percent interest of Argyle's junior partner. The bid failed. Rio Tinto, the senior partner, bought Ashton instead, consolidating Rio Tinto's diamond position.

◆

The center of the Indian diamond trade is Prasad Chambers, an ugly concrete tower that rises out of the dust and hubbub of Bombay's Opera House district. There is no opera anymore, but instead the massed traders and runners of the diamond quarter. The runners wear loose garments to conceal the rough taped to their bodies. The traders huddle in groups, switching parcels of diamonds. Prasad Chambers stands amid the press. Streams of people file past the guardhouse and into the dirty courtyard, and from the courtyard into the lobby. They queue for the tiny elevators in long lines that wind around the lobby and out the door. Although the building has been occupied for years, it remains unfinished. Debris litters the stairwells. In the corridors, bare bulbs hang from exposed wires. Lightbulbs cast a feeble light into the dark halls. Dirt lies piled in corners and the air smells of cement. Cement flakes from the walls and ceiling. Guards with rifles perch on battered chairs. Behind the locked doors of the diamond offices lies a different world, the world of the marble desk, of the ceaseless calls to Antwerp, Tokyo, Taipei, Tel Aviv, New York.

Eighty percent of the world's gemstone diamonds move through companies controlled from Prasad Chambers. What with the grubby appearance and the crowd surging all around, it seems calamitous, and perhaps it is this helter-skelter look that promotes the opinion, heard in the older cutting centers, that Indians don't love diamonds, only money. Arun Mehta smiled at this in his office on the top floor of Prasad Chambers, with a billion dollars a year in rough moving through his companies. "Well," he said, "perhaps we are aggressive."

One day I drove out along the seafront and up into the green precincts of the Malabar Hill. Old mansions lay behind crumbling walls, with sometimes only a gable or a curved roof visible among trees. On Cliff Road, I pulled in behind Russell Mehta's red Mercedes-Benz sports car and made my way through the building site of his new house to meet him. Lumber lay all around in piles, and an artisan with chisels and gouging tools sat with a length of dark wood across his lap, carving a pattern found by Russell Mehta's wife. Masons were finishing the marble terrace, where fountains would splash into basins and where guests would stroll to the balustrade at the brow of the hill and gaze out at the Arabian Sea.

Mehta and I climbed up through the rising mansion to the top floor, where his own bedroom would open to a private terrace with a still more breathtaking view of the sparkling sea. But the house on the Malabar Hill was a symbol of more than the rise of a family. It signified the emergence of a new order of diamond jewel, easily available to people who could not otherwise afford to buy diamonds. The success of this new trade, swelling the market with polished diamonds in the hundreds of millions, has altered the calculations of those who search for mines. A diamond pipe that might not have been profitable thirty years ago may be packed with tiny diamonds eagerly sought today by the India trade. The demand created in Bombay and Surat was part of the mathematics of discovery that rippled out to shores far distant from those of India.

12

The Dogrib Country

The appetite for diamonds is a powerful hunger, and it transforms the places where it is awakened. When the three garimpeiros found the big pink in the Rio Abaete on that May morning, it sparked the immigration of thousands of garimpeiros into the region from elsewhere in Brazil, bringing a time of lawlessness to the diamond rivers. In the Barrens, too, discovery provoked a rush of newcomers. A vast and little-traveled land suddenly reverberated with the noise of helicopters. If the strike in the Barrens shook the old order of the diamond world, so diamonds shook the balance in the Barrens. A restless wind is blowing through the diamond realm today, and power is shifting from hand to hand. New players have arrived to stake their own claims in the diamond business, sometimes displaying an unsuspected aptitude for the game. One of these was the

Dogrib Indians, who faced down one of the biggest mining companies in the world.

In March 2000, on the ice road from Yellowknife, convoys bound for the diamond camps were going day and night. Massive trucks crept gingerly onto the ice and began the eighteen-hour journey from lake to lake. At the yards in town where the trucks formed up, the urgency was tangible. A single menace hung over the enterprise, as it did every year—spring.

From Yellowknife to the eastern end of Lac de Gras, the ice road traverses 280 miles. It crosses lakes where the ice in March is six feet thick, easily strong enough for the heaviest rigs. Yet as the days lengthened, the sun attacked the road at its vulnerable points. On the land portages between the lakes, patches of exposed dirt attracted the sunlight, and frozen ground could melt into impassable muskeg in hours. The road was open to heavy trucks for a scant six weeks. Through that narrow window, remote mine sites had to push most of the material they needed for the whole year. These were the facts faced by Rio Tinto that spring as it waited for the last of the government permits needed to complete the trucking of construction materials north to its growing mine site at Eira Thomas's discovery, now named Diavik.

Diavik's managers thought they had cleared the last environmental hurdle, when suddenly federal officials denied a crucial permit. Rio Tinto and Aber Resources had by then spent five years sampling the pipes and mapping the extent of the targets to be mined. They had commissioned environmental studies of the lake. They had spent some hundreds of millions of dollars and submitted thousands of pages of documents. Now, without warning, the government cited an environmental technicality and blocked their trucks from the road.

In a rage, Rio Tinto and Aber shut down the Diavik site and laid off staff. In Yellowknife, where a gold mine had recently shut down, the fear grew that the diamond mine would be canceled outright. Because there are no year-round roads into the Barrens (the cost of construction being prohibitive), losing the trucking window of the winter road would cause a delay of one full year. A year's delay would cost Rio Tinto $50 million in interest on the project's debt, it was claimed, and industry watchers speculated that the company might shelve the mine indefinitely rather than suffer the charge.

As it turned out, the immediate fate of the Diavik site did not lie in the hands of the federal government's bureaucrats, who were denying Diavik the necessary permits, but with the Dogrib. Lac de Gras lay in the ancestral lands of several native peoples, including the Dogrib, and Dogrib tacticians had determined to hold up the permits until they were satisfied with the compensation package Diavik was offering their people. The Dogrib intervention represented a revolutionary shift of power. A small and largely unlettered native group, spread through the bush at the edge of the Barrens, had halted, literally in its tracks, the development of a C$1.3 billion mine that was expected to employ 450 people for twenty years.

◆

The Dogrib capital, Rae, lies sixty miles from Yellowknife, on a height of land above Marian Lake. The silver steeple of St. Michael's Catholic church gleams against the sky. A winter road runs north up Marian Lake, connecting to the distant Dogrib communities at Lac La Martre and Rae Lakes. In a territory the size of Scotland live some three thousand Dogrib, seventeen hundred of them in Rae.

The attachment of the Dogrib to fine distinctions of place is reflected in the old pattern of habitation within Rae. One day John B. Zoe, a forty-two-year-old Dogrib and the chief negotiator for his people in the talks with Diavik, drove me along a snow-packed road beside the lake. He identified the different Dogrib clans who lived there, organized by kinship according to unmarked borders that were clearly understood. "These are the Tagah Goti, the Follow-the-Shore Dogrib," Zoe said as we passed a few houses. "Their family names are Rabesca, Wellin, and Blackduck. Now there's a fine line right here. Now these people," he pointed to the next houses, "these are the Chocolates and the Huskys. They are Etati, the People Between Other People, because they come from between here and Great Bear Lake."

And so it went—the Martin Lake People, the People from the Edge of the Woods. Now diamond money has started changing this, eroding the old groupings by kinship and place. In the thin bush at the edge of Rae, planners have laid out new streets. Satellite dishes crane at the sky beside modern, suburban houses with aluminum siding and barbecues on the decks. The families living in this new part of town live here not because their relatives do but because they can afford it. The new trucks parked outside reflect this status and signal a new means of differentiation—money. For some of the Dogrib have well-paid jobs at the BHP mine, which opened in 1998.

But the Dogrib extracted more than money from the diamonds. They took power, too, and in early 2000 that newfound power stood between the Diavik partnership and its deposit of 107 million carats of good, white diamonds sitting under Lac de Gras.

The natives' ability to halt the mine was proof of how far the world had come from the old days of the diamond empire. In their transactions with the diamond companies, the Dogrib effectively exchanged their past for a stake in the present. They did this relatively quickly, helped by a trend toward the empowerment of aboriginal peoples that, at the time of the diamond discovery in 1991, had already created a native political class.

When the diamond news broke, the Dogrib leadership saw the opportunity immediately, but realized they had no legal right to influence the headlong commercial events. "It was such a mad rush," said Zoe. "I remember what happened at Wekweti. About a hundred people live there, maybe a dozen families. There's a hill behind it, and a helicopter landed on the hill. The community wasn't on any maps, and the Dogrib saw these guys come out of the bush with stakes, and they drove staking posts right down the middle of town. These people had a mandate to stake, and they staked. We were just starting to negotiate our land claim, and the diamonds added urgency to the claim. It showed us we don't have protection from anything except what the government deems is ours at their whim and will. At that time, when they found the diamonds, we didn't have more than a few Dogrib working in all the mines [in the Northwest Territories]. So we decided to get some benefits."

The Dogrib evolved a legal position that rested on their origins as a hunting people. To demonstrate those origins, the chiefs sent young Dogrib researchers out into the far-flung communities to gather the hunters' tales. The stories they collected showed how closely the traditional livelihood of the people depended upon knowledge of the land.

> However the animal roams around, it don't usually go back the same way. Even our ancestors say that. . . . Like out in Wekweti near where they call Kwiakw'ati. And it [the caribou] goes straight to Nodiik'e, we know that. And when it has to travel back to Barrenland it goes all the way back on the other side of Snare Lake called Ts'inazee, back to Barrenland. [Johnny Eyakfwo, age seventy-three]

> The caribou used to travel past Behtsoko, and towards Yahtideeko as far as to a place called K'iti and along Edazi, and once they settled in the area, if the food is plentiful, they lived and ate there for a long time. [Joe Zoe Fish, age seventy]

The place-names in the stories, then, were the material assets of a hunting culture—the essential data for survival. It followed in the Dogrib view that the destruction of such names by any means, such as altering the topography to dig a mine, constituted the destruction of assets, and someone must pay. BHP readily agreed, and in 1994 the Australian miner paid C$1.2 million into the Dogrib treasury, and made formal commitments to employ aboriginals. When Diavik (Rio Tinto and Aber) came to negotiate, the price is said to have been higher. Diavik balked,

Tankers and loads of heavy equipment for the diamond camps backed up in Yellowknife waiting for permission to travel the winter road. (Matthew Hart)

and that is why the miner's trucks did not go up the winter road.

In the end, the Dogrib reached a settlement with the Diavik joint venture. The Yellowknives, Chipewyans, Inuit, and Metis also made pacts with the miner. Even the old emperor himself, De Beers, has sat down at the table with the sons of illiterate hunters and fishermen. De Beers now owns a diamond-bearing kimberlite dyke at Snap Lake southeast of Lac de Gras, and has found a large diamond pipe west of Attawapiskat, an Indian community at the bottom of Hudson Bay. It has entered into agreements with the native peoples in both places. What a far cry from when the modern diamond age began with a charge across the veld. No one had stopped to consult the Griqua. The imperialist spirit had flourished then, and out of that historical moment came the diamond cartel; but that was the nineteenth century, and this the twenty-first.

◆

Now in the long winter months, when summer's daylight has fled away, the great camps at Lac de Gras are visible from the air at a distance of twenty miles, blazing like diamonds on the breast of night. The BHP mine produces 4 million carats a year, and Eira Thomas's pipes under Lac de Gras will add another 6 million carats to that, so that some 15 percent by value of the world's gem diamonds will come out of those first two mines alone. Much of the production will proceed to market through channels outside De Beers. Only 35 percent of BHP's rough now goes to London to be sold. Aber Diamond Corporation, as it became, will receive 40 percent of the Diavik production as its own, or 2.5 million carats a year. In a deal with Tiffany & Co., the greatest diamond merchant in New York, the choicest of Aber's goods will be sold directly to the retailer, bypassing the DTC by a country mile.

As all this was unfolding, De Beers fought and lost its first Australian battle, the war of the cheap rough, then fought and lost its second, the attempt to buy a stake in Argyle out from under the nose of Rio Tinto. Even more ominously for De Beers, the Russians, with 20 percent of world rough, were said to be restive, and the possibility grew that they would finally jettison their long, uneasy partnership with the South Africans and market their goods themselves. In that event, De Beers's share of the rough market would have plunged from 80 percent to less than 50 percent in a decade. It was as if the spirit of the day was against De Beers, and time itself had come for the old cartel.

Yet De Beers was not in disarray. A powerful group of executives was preparing a stroke bold enough to have come from the mind of Sir Ernest himself. On February 1, 2001, the diamond company stunned the industry with the announcement that plans were afoot to take De Beers private—delisting its stock, removing it from the scrutiny of public ownership, and consolidating the power of the Oppenheimer family. The deal was structured as a buyout of De Beers by a consortium held 45 percent by the Oppenheimers, 45 percent by Anglo American, and 10 percent by Debswana, the De Beers-Botswana joint venture that owns De Beers's most important group of mines. The buyout upended De Beers's most recent strategy, that of boosting the share price of the company and attracting a wider ownership through such tactics as increased openness about the company's affairs, jettisoning the London stockpile, and abandoning the role of rough-market manager. Alas, the share price had not improved, even in the face of record diamond sales. Nicky Oppenheimer apparently concluded that the stock was effectively on sale, and if no one else wanted it he would buy the $18-billion company himself.

As the great diamond company vanishes from the public listings, it will not leave accountability behind. The old diamond world that De Beers created is no more; the cartel is a footprint filling in the sand. But there is no power in diamonds to match the powert of De Beers, and the leadership of the diamond world is a mantle that De Beers cannot shed. At this juncture in the history of diamonds, when the repute of the jewel is suffering from its connection to a wicked trade, that cloak of leadership may be a heavy garment.

More than seventy NGOs are ranged against the war trade in diamonds. In response to their efforts, Antwerp's Diamond High Council has sent inspectors into cities in Sierra Leone and Angola to certify that the goods coming from there are "official" production, and not from war. But are such warranties reliable? The goods could easily be contaminated by war rough before they reach the certifying inspector. Moreover, the main war production from these countries does not come marching boldly through the capital cities but flies directly from the rebel areas where it is mined.

Neither is there much progress toward providing a warranty chain for legitimate goods. De Beers spokesmen have suggested that the notices now inserted into sight boxes, stating that the goods are free of faint, provide an assurance that shelters the clients downstream. But the clients buy rough elsewhere too, and could use their De Beers warranty as a cover for any goods they polished, since no subsequent customer—the polisher's customer—could know whether the original rough had come from De Beers in London or arrived in Antwerp in the dead of night in the sole of someone's shoe.

Much diamond business proceeds behind a screen of spurious attributions. Masses of goods are regularly sluiced through ports of convenience, where they pick up the commercial equivalent of phony passports. Switzerland is not a diamond-mining country, yet because dealers channel thousands of carats through Swiss free-trade zones (*Freiläger*), Switzerland is the listed country of origin for large parcels of goods. It is the same in Gambia, a tiny west African nation with no diamond production of its own, where diamond-trading companies acquire illicit rough and pack it off to Antwerp and Tel Aviv as "Gambian." Even in Canada, the apple-cheeked freshman of the diamond world, federal police have warned that criminals may be mixing war rough with arctic goods.

Of course, no special criminal class is needed. In April 2001 the Belgian newspaper *Le Soir* reported that Belgian military intelligence had found large Antwerp firms dealing in UNITA war goods. A panel of experts reporting on the war-diamond trade to the United Nations Security Council drew a picture of an intractable commerce untroubled by censure. No doubt the diamond trade knows that the NGOs active in the campaign against war diamonds are reluctant to espouse a consumer boycott of diamonds for fear of hurting Africans employed in legitimate mines. But the feeling is growin that the preservation of one African's job is not worth the cost of another African's life.

No one can cripple the war trade as effectively as De Beers. When it ordered its buyers to halt all purchases of loose rough coming into the trading cities, it dealt a swift, hard blow to war traders, for it took as much as $15 million a week straight out of the market. Certainly other buyers moved in to fill the vacuum, but it is still true that it became more difficult to move the goods in the volumes necessary to sustain war.

Some diamonds now come to market with reliable guarantees of origin. If public sentiment against the war trade begins to threaten sales, the demand for such "clean" gems could swiftly rise. Gary Ralfe, the De Beers chief executive, has admitted that in such circumstances De Beers might require clients of the DTC to keep De Beers goods separate from other goods, so that a clear trail from legitimate mine to market could be reliably affirmed. Even as an outside chance, the possibility of such a move would send shivers into the trade, because it would cast a huge mass of unsheltered goods into a suspect, second-class state, sharply lowering the value.

Assurances of diamond legitimacy could also come from a regime of independent inspection, such as that contemplated in draft legislation in the United States Congress in early 2001—the Clean Diamonds Act. If enacted, the bill would bar from the United States any diamonds unprotected by acceptable guarantees. The law would get its teeth from a right of inspection along the export chain. If inspectors were armed with some dependable means of detecting the origin of rough—say, Professor Rossman's method—any dealer prepared to handle war goods would have to accept the parallel risk of detection and possible ruin. War goods may account for only a small percentage of the world's rough, yet their power to harm the diamond trade is great. The old, secretive habits of the trade served the jewel well, but today is a different day, and with 115 million carats a year coming out of the ground, the diamond is a different jewel.

Many new players populate the diamond world today. Until the diamond strike in the Barrens, the only form of carbon mined by BHP was coal. Rio Tinto is today a much more powerful diamond miner than it was. But the Canadian diamond strike did more than enlarge the prospects of a few big companies. It catapulted intrepid characters into the diamond scene, many of them still there.

Since he drilled the Point Lake pipe, Ed Schiller has ranged the world looking for diamonds. He pops up at geology conferences, an irrepressible figure always boiling with passion for a new target somewhere in the world. Chris Jennings was toppled from the leadership of SouthernEra by dissatisfied shareholders after the Marsfontein debacle; later, in a countercoup, he seized control again. He divides his time between Canada and South Africa, and between platinum and diamonds.

Robert Gannicott is chief executive of Aber Diamond Corporation. Gren and Eira Thomas left Aber and started Navigator Exploration Corp. Their prospects include a joint venture with Canabrava Diamond Corporation, a junior run by Rory Moore, the South African who, with John Gurney, sounded the trumpet for the first Canadian strike. Now Eira Thomas and Moore are camped on a block of claims in the swampy lowlands of northern Ontario, adjacent to a thirty-acre kimberlite pipe held by a mining company that has become one of the most active senior miners in Canada—De Beers Consolidated Mines.

In early 2001 there was another diamond rush in Canada—this time in Manitoba.

Eira Thomas's former tent mate, Leni Keough, is still searching for diamonds, but will say where.

In February 2001, at Saxondrift on the Orange River, on the diamond ground brought into Trans Hex by Tokyo Sexwale, miners found a 216-carat diamond. It was frosted on the surface, and there was talk of polishing in a window so prospective buyers could take a closer look. In the end they decided not to, and sold it as it was for some $850,000. This was good not only for Trans Hex but, as it turned out later, also for DiamondWorks, the little company that had been flattened in Angola. DiamondWorks had acquired new owners and bought a property of its own on the Orange River, right beside the ground where the big stone was found—as they soon pointed out.

Gabi Tolkowsky helped develop software that translates the unique refractive pattern of a polished diamond into music.

In diamond mines from Botswana to the Diamond Coast, from the Arctic Circle to Brazil, someone, somehow, is stealing rough.

◆

Diamonds are the dark and the light. They are windows polished into the heart of man. They have existed in the universe since before the Earth and the Sun. Those most ancient diamonds were showered into space in the inconceivable cyclones of exploding stars. In our own planet, at depths of a hundred miles, carbon formed into diamond crystals, which are both impervious and frail. Many diamonds abandoned their form and vanished into graphite on their journey to the light. Something of that ineffable fragility remains present in the jewel. When the Persian conqueror Nadir Shah overran Delhi, he ransacked the mogul's palace for the most famous diamond of the world until finally one of the harem women revealed that the emperor kept the jewel hidden in his turban. Nadir Shah invited the vanquished ruler to a feast, where, seizing upon an Oriental custom, he offered an exchange of turbans. The mogul could not refuse. Nadir Shah calmly took the proffered turban and put it on his head. Later, he rushed back to his private quarters, pulled the turban apart, and found the diamond. "Koh-i-Noor!" he is said to have gasped, giving the diamond its name: Mountain of Light.

The diamond trade unpackages and sells this light. No doubt they are a bit corrupted by it. A mound of clean rough is an intoxicating heap: It throbs with potential light. If you plunge in your hands the stones feel like silk; they slide through your fingers with a hiss. They come from the hammerstroke of creation. The diamond cutter must interrogate the visible record of a diamond's past, then put the diamond to the wheel. Somewhere now the big stone that the garimpeiros suctioned from the Abaete on a hot May morning has been transformed into a tango of pink light. Or else it shattered into dust. That's the great adventure.

Appendix

Key to Abbreviations

BHP—The Broken Hill Proprietary Company: an Australian mining conglomerate with diamond interests in Canada.

CSO—Central Selling Organization: now abandoned in favor of the appellation DTC.

DTC—The Diamond Trading Company: De Beers's selling arm.

UNITA—The National Union for the Total Independence of Angola: Angola's rebel faction, active in diamond mining.

MPLA—The Popular Movement for the Liberation of Angola: Angola's governing party.

NGO—Non-Governmental Organization: the term for activist and humanitarian groups not officially allied with governments although often dependent on them for funds.

Selected Bibliography

Balfour, Ian. *Famous Diamonds*, 2nd ed. Santa Monica: Gemmological Institute of America, 1992.

Blakey, George. *The Diamonds*. London: Paddington Press Ltd., 1977.

Bruton, Eric. *Diamonds*, 3rd ed. London: N.A.G. Press, 1976.

Capon, Tim. "The Role of De Beers: Past, Present, Future." *Mazal UBracha Diamonds* (Tel Aviv). 1996.

De Beers Consolidated Mines Ltd. *Diamond Security and Illicit Diamond Trafficking*. Johannesburg: Paper prepared by De Beers Group Executive Diamond Security, n.d.

De Beers Consolidated Mines Ltd. *Notable Diamonds of the World*. Undated monograph.

De Beers Consolidated Mines Ltd. *The Centenary Diamond*. Undated monograph.

De Beers Consolidated Mines Ltd. and De Beers Centenary AG. Circular to Holders of Linked Units [disclosure document to shareholders prior to buyout of De Beers by Oppenheimer et al.] 2001.

De Beers Consolidated Mines Ltd. and De Beers Centenary AG. Annual Reports 1999.

De Beers Consolidated Mines Ltd. and De Beers Centenary AG. Annual Reports 1998.

De Boeck, Filip. "Domesticating Diamonds and Dollars: Identity, Expenditure and Sharing in Southwestern Zaire." *Development and Change,* Vol. 29, no. 4. October 1998.

Duval, David; Green, Timothy; and Louthean, Ross. *The Mining Revolution: New Frontiers in Diamonds.* London: Rosendale Press, 1996.

Even-Zohar, Chaim. "Global Witness May Have Forfeited the Right to Walk the Moral High Road." *Mazal UBracha Diamonds* (Tel Aviv). Vol. 15, no. 115.

Fipke, C. E.; Gurney, J. J.; and Moore, R. O. *Diamond Exploration Techniques Emphasising Indicator Mineral Geochemistry and Canadian Examples.* Geological Survey of Canada, Bulletin 423 (1995).

Frolick, Vernon. *Fire into Ice: Charles Fipke & the Great Diamond Hunt.* Vancouver: Raincoast Books, 1999.

Fumoleau, Rene. *As Long as This Land Shall Last: A History of Treaty 8 and Treaty 11.* Toronto: McClelland and Stewart Ltd., 1973.

Gibson, Roy. "Argyle Papers Kept Secret, Says Lawyer." *The West Australian* (Perth). Oct. 12, 1998.

Gooch, Charmian, and Yearsley, Alex. *A Rough Trade.* London: Global Witness Ltd., 1998.

Gosman, Keith. "Diamond Heist." *The Sydney Morning Herald* (Sydney). Apr. 16, 1994.

Government Diamond Valuator, Republic of South Africa. Presentation to Acting Director General. Undated internal briefing document.

Government of the Northwest Territories. *Diamonds and the Northwest Territories, Canada.* Yellowknife: 1993.

Haggerty, Stephen E. "A Diamond Trilogy: Superplumes, Supercontinents, and Supernovae." *Science* 285: 851–60.

Harlow, George E. "What Is Diamond?" In: *The Nature of Diamonds* (Harlow, George E., ed.) Cambridge and New York: Cambridge University Press and American Museum of Natural History, 1998.

———. "Following the History of Diamonds." In: *The Nature of*

Diamonds (Harlow, George E., ed.) Cambridge and New York: Cambridge University Press and American Museum of Natural History, 1998.

Hughes, Judy. "Argyle Inquiry Attacks Police." *The Australian* (Sydney). Sept. 6, 1996.

Isenberg, David. *Soldiers of Fortune Ltd.: A Profile of Today's Private Sector Mercenary Firms.* Washington, D.C.: Center for Defense Information, 1997.

Jackson, Stanley. *The Great Barnato.* London: Heinemann.

Jennings, C. M. H. "The Discovery of Diamonds in Botswana." *South African Journal of Science.* Vol. 66, no. 8.

Jessup, Edward. *Ernest Oppenheimer: A Study in Power.* London: Rex Collings Ltd., 1979.

Kaplan, David E. and Caryl, Christian. "The Looting of Russia." *U.S. News & World Report.* Aug. 3, 1998.

Khalidi, Omar. *Romance of the Golconda Diamonds.* Middletown, N.J.: Grantha Corporation, 1999.

Kirkley, Melissa. "The Origin of Diamonds: Earth Process." In: *The Nature of Diamonds.* (Harlow, George E., ed.) Cambridge and New York: Cambridge University Press and American Museum of Natural History, 1998.

Kirkley, Melissa B.; Gurney, John J.; and Levinson, Alfred A. "Age, Origin, and Emplacement of Diamonds: Scientific Advances in the Last Decade." *Gems & Gemology.* Spring 1991.

Kramer, Andrew. "Four Convicted in Russian Diamond Embezzlement Case." *The Associated Press.* Moscow, May 17, 2001.

Krashes, Laurence S. (Ronald Winston, ed.) *Harry Winston: The Ultimate Jeweler.* Harry Winston Inc. and the Gemmological Institute of America. New York and Santa Monica: 1984.

Legat, Allice. *Caribou Migration and the State of the Habitat: Annual Report.* Yellowknife: West Kitikmeot Slave Study Society, 1998.

Newman, Peter C. *Company of Adventurers.* Toronto: Viking, 1985.

Pearson, Carl. *A Walk on the Demand Side*—Revitalize Demand. London: Unit Publications. Presentation notes for an address to a business conference in Cape Town, 2000.

Reardon, David. "Police Urge Open Inquiry into $2 Million Argyle Diamond Theft." *The Sydney Morning Herald* (Sydney). Oct. 15, 1998.

Rist, Curtis. "Neptune Rising." *Discovery.* Vol. 21, no. 9: 54–59.

Rubin, Elizabeth. "An Army of One's Own." *Harpers*. February, 1997.

Saxon, Martin. "Dozens in Diamond Ring." *The Sunday Times* (Perth). Sept. 10, 1995.

Skidmore, Thomas E. *Brazil: Five Centuries of Change*. New York: Oxford University Press, 1999.

Smillie, Ian; Gberie, Lansana; and Hazleton, Ralph. *The Heart of the Matter: Sierra Leone, Diamonds & Human Security*. Ottawa: Partnership Africa Canada, 2000.

Spiegel, Maura. "Hollywood Loves Diamonds." *The Nature of Diamonds* (Harlow, George E., ed.) Cambridge and New York: Cambridge University Press and American Museum of Natural History, 1998.

Surowiecki, James. "The Diamond Market vs. the Free Market." *The New Yorker.* July 31, 2000.

Treadgold, Tim. "Argyle's Bitter Diamond War." *Business Review Weekly*. Nov. 4, 1996.

Wharton-Tigar, Edward (with Wilson, A. J.) *Burning Bright: The Autobiography of Edward Wharton-Tigar.* London: Metal Bulletin Books Ltd., 1987.

Williams, Roger. *King of the Sea Diamonds: The Saga of Sam Collins.* Cape Town: W. J. Flesch and Partners, 1996.

Note on Additional Sources

Some financial houses produce excellent diamond-industry reports, and I am indebted to the analysis of such writers as James Allan of Barnard Jacobs Mellet, Johannesburg, and Jack Jones of CIBC World Markets, London. My reconstruction of the Canadian diamond rush relied on scores of newspaper accounts, including reports in *The Globe and Mail, The Financial Post,* and *The Vancouver Sun.*

To stay current with developments in the diamond wars I regularly checked *The Angola Peace Monitor* on the Internet at *www.anc.org.za/angola* and the correspondence and reports of two United Nations groups: the Monitoring Mechanism on Sanctions against UNITA, and the Panel of Experts on Sierra Leone Diamonds and Arms. The work of both UN bodies may be followed by navigating from *www.un.org/documents/scinfo.htm.*

Acknowledgments

First I must thank the explorer Chris Jennings for his untiring help with the story of discovery in Botswana and Canada. The geologist Ed Schiller offered valuable advice, and Stuart Blusson added details to the account of exploration. Hugo Dummett, Robert Gannicott, and Gren Thomas all read parts of the text, to its benefit. Eira Thomas, Robin Hopkins, and Leni Keough also read and corrected sections. I am grateful to Stephen Haggerty at the University of Massachusetts for his kind help with the account of diamonds in space. John Gurney examined and corrected passages describing his work on diamond indicator minerals, and Rory Moore and the indefatigable George Read labored hard to teach me the lessons of the upper mantle. Vern Rampton made suggestions for improving the account of glaciation. George Harlow of the American Museum of Natural History, an accomplished writer on diamonds, improved the description of the atomic structure of diamond. If I have mangled anything in the service of brevity or plainness, the fault is mine.

Thanks to Randy Turner whose pilots flew me into his tent camp in the Barrens, where the diamond geologists Melissa Kirkley, Nik Pokhilenko, and Jennifer Irwin showed me how to dig for indicators. Thanks also to Graham Nicholls at BHP, and to Bill Trenaman for his help with travel to Angola. For help with the account of Golden ADA, thanks to Joe Davidson at the FBI, to the San Francisco investigator Jack Immendorf, and to David Kaplan at *U.S. News & World Report.* For scrutinizing passages about the cutting of large stones at William Goldberg Diamond Corporation, and for other help, thanks to Barry Berg. I am also grateful to Simon Teakle and to that sparkling gentleman, the Manhattan gem dealer Richard Buonomo.

Ian Smillie read draft text on war diamonds and made useful suggestions. For my understanding of the mechanics of the diamond wars, special thanks to Jakkie Potgieter at the Institute for Security Studies in Pretoria and Ed Dosman of the Centre for International and Security Studies at York University, Toronto. My gratitude to Stephen Fabian, to André Louw, the sea-diamond miner, to Niel Hoogenhout, and Peter Danchin. Thanks to the Mehta family of Bombay and Antwerp, to Sunil Shrivastava, who showed me the best way to hold a loupe, and to the family of Piyush Shah, who were kind to me in Surat. Gerald Rothschild and Mark Boston reviewed a list of assertions about India, increasing my comfort in making them.

Tim Treadgold in Perth made good suggestions about the dispute between Argyle and De Beers, and also helped to verify my account of the theft of Argyle's pinks. Thanks to Diana Bagnall in Sydney for the many clippings about the Argyle thefts, and special thanks to the news librarians at *The West Australian* and *The Sunday Times*, both of Perth, who helped to verify names and dates in my account. Davy Lapa was generous with his insights on the Antwerp trade. Carl Pearson and Katharyn Barnett know everyone in diamonds, and helped me hugely. Charles Wyndham read and corrected the passage on theft at the DTC. Enormous thanks to Richard Wake-Walker, whose diamond knowledge is encyclopedic, who offered many valuable suggestions, and who also allowed me to quote from his company's report on the large Brazilian pink.

Many thanks to David Blotner, assistant chief of criminal litigation in the antitrust division of the United States Department of Justice, who reviewed certain statements in the book.

I owe much thanks to De Beers for assistance that goes back several years, beginning with George Burne, now retired from the board, a diamond mandarin of the old school. Roger van Eeghen in London dealt gracefully with a stream of e-mails that flowed upon him for two straight years. Thanks also to Tom Beardmore-Gray, Chris Welbourn, Joe Joyce, Tracey Peterson, Tom Tweedy, Rory More O'Ferrall, Chris Alderman, Andrew Lamont, and Gavin Beevers, an executive in Botswana when I met him and now head of operations in Johannesburg. I am indebted to Sir Alan Grose for discussing diamond security, and to De Beers's chief executive, the mannerly Gary Ralfe, for grace under fire. Thanks too to Nicky Oppenheimer.

Thanks to *The Atlantic Monthly*, where my account of theft on the Diamond Coast first appeared, and especially to my editor at the magazine, Amy Meeker. I owe much to Stephanie Wood, Susan Walker, David Hull, Jefferson Lewis, and Alex Beam. Neal Bascomb helped to plan the book. George Gibson and Clive Priddle offered good advice on the manuscript. Were it not for my agent, Michael Carlisle, there would be no book at all. I owe enormous gratitude and affection to Jackie Johnson, my editor. My deepest thanks are to Heather Abbott, who saw the work and the writer all the way through.

Index

A

A–21, 104, 105, 108, 109
A–154, 109, 110
Abdul Hamid II, 158
Aber Diamond Corporation, 247, 250–51
Aber Resources, 93–94, 95, 96, 97, 102, 103, 109, 242
  and Dogrib, 245–46
  locating targets, 99–100
  merger with DHK, 104–5, 107, 108
Advertising, 138–42, 198
Africa, 33–37
  and war diamonds, 183–92
African National Congress (ANC), 114–15
Afrikaners, 118
Agra diamond, 153
Albert, Prince, 147
Alberta, 119
Alexander Bay, 161, 166
Alexander the Great, 144
Alexkor, 160, 161, 168–69
Alluvials, 5–6, 33, 52, 116, 118, 160, 182
Amsterdam, 48
Anamint, 123
Anglo American Corporation of South Africa, 29, 49, 50, 114, 123, 136, 247
Angola, 84, 120, 197
  war diamonds, 183–84, 188–89, 190–91, 193, 195, 248
Antwerp, 54, 84, 122, 123, 126, 135, 136, 172, 194, 203, 225, 235

cutters/cutting, 48, 222, 224, 229, 232
Indian cutters in, 230
Russian rough in, 180
secondary trade in rough, 233
switching in, 128–29
and war diamonds, 183, 187, 188, 192, 193, 249
Aquarium, 171
Argyle mine, 84, 171–73, 236–38, 247
Arthur Andersen (co.), 178
Ashton Mining Ltd., 236, 238
Asscher, Joseph, 46*f*, 204
Athwa Lines, 229
Atlantic seabed, mining, 164–69, 167*f*
Attawapiskat, 246
Auction, 149–53
Audit trail, 197
Australia, 84, 171–73, 236–38

B

B. Arunkumar & Co., 232, 233
Babur, 145, 153
Baburnama, 145
Bankers' Booklet, 237
Barker, Lee, 97
Barnato, Barney, 37–40, 38*f*, 41, 42–44, 49, 50, 53, 116
Barnato, Harry, 37, 38, 40
Barnato Mining Company, 40–41
Barrens, 73, 75, 197, 242, 243
diamond exploration camp, 96*f*
diamond production, 113
diamond strike in, 86, 87–88, 95, 102, 199, 241, 250
indicator minerals in, 76
rush in, 84–112, 118–19
winter tent camp, 89–90, 91–92
Beach mines, 161–62
Bechuanaland Protectorate, 55
Beit, Sir Alfred, 45, 48
Belgian Congo, 52
Benguela Concessions (Benco), 117
Benguela Current, 169
Berg, Barry, 203–4, 205–6, 209, 210, 212, 213–14
Berstein, Motti, 204–7, 205*f*, 213
Bhansali, Bhanuchandra, 231
BHP Minerals, 80, 82, 83, 85, 91, 93, 95, 97, 98–99, 102, 104, 113, 189, 250
camp, 103
mine, 244, 247
payment to Dogrib, 245
Big Hole, 41–42
Black Swan Resources, 4, 6, 10, 15, 19, 20–21
Blacks, South Africa, 115–16, 118
Blue Diamond of the Crown
see French Blue
Blue ground, 40
Blusson, Stuart, 70–72, 73, 75–76
Boers, 34, 35–36
Bombay, 122, 135, 223, 232, 238, 240
cutters in, 222, 224, 228–29
and war diamonds, 183
Botha, Louis, 50–51
Botswana, 33, 55–59, 56*f*, 61, 63, 65, 84, 102, 170–71
diamond production, 58
Boxes, 53, 131, 132, 133, 135, 193, 203, 229, 235, 248
Braganza stone, 17–19
Branded diamonds, 198–99
Brazil, 10, 17, 156, 203, 230
diamond production, 3, 5–6, 33, 37
exploration in, 119
Brilliance, 230
cutting for, 211
Brilliant cut, 216, 217*f*, 227
British Columbia, 70
Broken Hill Proprietary Company, 80
Brokers, licensed, 131–32, 133
Brown diamonds, 3, 227, 236
Bryan, Doug, 81, 89, 90–91
Bubbles, 216–17, 219
Bultfontein, 36, 39

Burning Bright (Wharton-Tigar), 125–27
Bychkov, Yevgeni, 175, 179–80

C

Camafuca kimberlite pipe, 120
Cambodia, 187, 189
Campos, Geraldo, 2, 4, 16, 17
Campos, Gilmar, 2–4, 5*f*, 6, 10, 11, 15
Campos, Gisnei, 2–3, 4, 15–16, 17
Campos brothers, 2–3, 6, 10, 14, 17, 20
Canabrava Diamond Corporation, 251
Canada, 62, 73, 118, 122, 134, 190
  De Beers and, 72, 119, 237, 246
  criminal activity in, 249
  determining origin of rough diamonds, 195
  diamond production, 102
  diamond strike, 85, 250, 251
  exploration in, 119
  government diamond valuator, 125
  mineral rights in, 79
  mines in, 136
  second diamond rush in, 251
Canadian Arctic Diamond, 197
Canadian Shield, 73
Cape of Good Hope, 33
Cape Town, 34, 44, 117
  Table Bay, 115
Carbon, 22, 24, 25–26, 250, 252
Carbon atom, electronic structure of, 24
Cartel, 29, 45, 53, 55, 235, 246
  Argyle and, 236–37
  check that started, 44*f*
  end of, 113–37, 248
  greatest diamond lode in history in, 60
  position in rough market, 54
  power of, 114
  rough to, 58–59
  in South Africa, 114–21
  Soviets signed with, 54
Cartier, 142, 148, 221
Cartier, Pierre, 158
Caustic fusion, 107
Centenary Diamond, 212, 215–21, 221*f*
  facet plan, 219*f*
Central Selling Organization (CSO), 123, 125, 126, 188
Chandragupta, 144
Charter House Street
  Number 2 and Number 17, 121–22, 123–24, 127–30, 133, 136, 141, 214
Chawchinahaw, Captain, 74
Cheap rough, 247
  India, 229, 230, 237, 238
Chicapa River, 184, 185
Chow Tai Fook, 21
Christie's, 149–53
Chrome, 27–28, 79
Chrome diopsides, 26, 30, 70, 77, 78, 80, 81
Chromites, 30, 67
City and West East Ltd., 54
Claims, 39–40
  in Barrens, 96, 99, 100–101, 102
  in Barrens: staking, 79, 81, 89, 90–91, 92–93, 95, 96, 101, 102–3
  in North America, 77, 79, 81, 83
  in Ontario, 251
Clarity, 124, 127, 211, 213
Clean Diamonds Act, 250
Cleavage plane, 204
Cleaving, 204, 205, 207–8, 217
Collins, Sam, 165
Color
  gradations of, 14, 128, 132, 213–14
  large pink, 3, 7, 9, 10, 16–17, 19
  sorting by, 124, 127
Colorado, 69, 119
Compagnie Française des Mines de Diamant du Cap de Bon

Espérance (French Company), 42
Consumer boycott (proposed), 192, 193
Coopers & Lybrand, 172–73
Coromandel, 4, 5, 8, 19
Corridor of Hope, 95
Côte-de-Bretagne, 156
Cratons, 31, 33, 69
Crimmins, Barry, 172–73
Crimmins, Lynette, 172–73
Crystals, 23, 24, 124, 132
  formation of, 212
CSO
  see Central Selling Organization (CSO)
Cullinan, Thomas, 45–46
Cullinan I, the Great Star of Africa, 46, 46*f*, 47, 47*f*, 204, 221
Cuts, 216
Cutter(s), 19, 200, 201–22, 252
  in India, 147, 154, 227, 229, 230, 231, 236
  Rosy Blue, 234
Cutting, 210–12, 217

D

Dalhousie, Lord, 146, 148
Danziger, David, 201–4
Davidson, Joe, 177, 178, 179
Davis, Geena, 143
De Beers, 28, 43, 44–45, 50, 51, 133–34, 148, 171, 212
  advertising, 138–42
  and Argyle mine, 236–38
  branded diamonds, 198–99
  in Canada, 72, 119, 237, 246
  Centenary Diamond, 214–21
  controlling supply/market, 63, 113, 114, 135–36
  on Diamond Coast, 160
  end of old cartel, 120, 136–37
  and India trade, 233
  and Kalahari diamond hunt, 63–64, 65, 66, 68
  management review, 136
  mining seabed, 164–69, 167*f*
  in North America, 71–72, 81, 90–91, 93, 95
  Oppenheimer elected chairman of, 53
  Oppenheimer's assault on, 51–53
  power of, 29, 47, 54, 248
  retail stores, 137
  rough diamond sales, 53, 84–85, 122, 123
  and Russian rough, 174–75
  self-dealing among directors, 48
  share of world production, 47
  sorting regime, 124–27, 131–32
  and Soviet challenge, 54–55
  and speculation in rough, 235–36
  taking private, 247–48
  thefts from, 163–64, 168, 181
  and war diamonds, 182, 189, 190, 193, 194, 249–50
de Beers brothers, 36–37
De Beers Consolidated Mines, 27, 251
De Beers Mining Company Ltd., 41
de Ville-d'Avray, Thierry, 154, 155
Deal Sweetener, 148
Debswana, 247
Demand
  for Argyle's diamonds, 238
  law of, 174
  for polished, 232
  for rough, 29
  for small diamonds, 240
  speculation and, 235
Democratic Republic of the Congo, 134, 186
DHK, 103, 104–5, 108
  collapse of, 112
Dhulip Singh, 145, 147–48
Dia Met Minerals, 76, 79, 83, 95, 102
Diamantaires, 7, 12, 208
Diamdel NV, 135

Diamond Area 1, 160–61, 163
Diamond Coast, 51, 84, 165
theft on, 160–64, 169–71
vacuuming rough, 162*f*
Diamond Control, 131
Diamond Corporation, 123, 125
Diamond Corporation of West Africa, 134
Diamond empire, 53, 54, 55, 59, 244
borders breached, 60
Diamond formation
minerals in, 27, 30
Diamond High Council, 187, 248
Diamond hunt
in Kalahari, 62–68
in North America, 68–83
in South Africa, 119–21
Diamond-hunting system, 27, 29–30
Diamond Research Laboratory, 215
Diamond rush
Barrens, 118–19
Canada, 251
South Africa, 34–37, 38
Diamond Security Organization, 182
Diamond stability field, 24, 25, 30, 79
Diamond trade, 11–12, 200, 208, 248–49, 252
low end of, 228
modern, 44
polished side of, 197–99
pricing conventions, 4
revolutionized in India, 230
and war diamonds, 183–92, 193–99, 249, 250
see also India trade
Diamond traders
Minas Gerais, 14–15
Diamond Trading
Company (DTC), 122–23, 129–30, 131, 132, 134, 193, 247, 249
sorters, 124*f*
Diamond wars, 182–200
Diamond world
capital of, 121
change in, 241
De Beers leader of, 248
modern, 33
new players in, 250–51
Oppenheimer family at head of, 116
rumor in, 113
Diamonds
anonymity of, 174, 198, 251–52
antiquity, 22–23, 32, 251–52
appetite for, 241–42
control of, 50
flaws in, 12–14, 210
great, 11–12, 221–22
greatest lode in history, 60
illicit, 180–81
impurities in, 195
large, 4, 11, 15, 18, 19
mystique of, 116, 144–49
search for, 55–59
sources of, 156
see also Rough
DiamondWorks, 184–85, 186, 251
Diatreme, 26
Diavik site, 242–43, 244–45, 247
Diggers' Republic, 35–36
Discovery block, 90, 95
Dogrib Indians, 242, 243–46
Doyle, Buddy, 110–11
DTC
see Diamond Trading Company (DTC)
du Plessis, Dawie, 220–21
Dummett, Hugo, 69, 70, 71, 72–73, 76, 80, 81–83, 90
Dunkelsbuhler, Anton, 48

E

Electromagnetic survey (Em), 100
Eclogite, 30, 72, 78
Eclogitic garnets, 72, 78–79
Eliason, Daniel, 157
Elizabeth, Queen, 145

Estrela Rosa do Milênio (Pink Star of the Millennium), 17, 19
Eureka (diamond), 34
Europeans in North America, 73–76
Exeter Lake, 77, 78, 95, 101
Eugénie, Empress, 149
Eugénie Blue, 149
Even-Zohar, Chaim, 199
Executive Outcomes, 184

F

Fabian, Stephen, 4, 6–7, 8, 12, 14, 15–17, 20, 21
Facets, 211, 227
  Centenary Diamond, 219, 219*f*, 220
Fakes, 15, 160
Falconbridge Limited, 61–62, 63, 64, 67–68, 77, 82
Fatal Transactions, 190
Federal Bureau of Investigation (FBI), 177–78, 179, 180, 181
Finland, 119
Finsch, 58
Fipke, Charles, 69–71, 73, 75–83, 90, 95, 96, 100, 101–2
Fipke's Curse, 95
First World War of Africa, 186
Fowler, Robert, 190, 194
Fraud, 174–81
French Blue, 154, 156–58
French Revolution, 156
Frostbite, 90

G

Gambia, 249
Gannicott, Robert, 86, 86*f*, 87, 94, 97, 119, 250–51
Garimpeiros, 1–4, 3*f*, 5, 6, 7*f*, 14, 15, 21, 23, 84, 241, 252
Garde-Meuble, 154–55, 157
Garnets, 26–27, 30, 31, 67
  from Barrens, 98, 104, 105
  and diamond prospectivity, 57
  G10, 28, 29, 30, 64–65, 69, 72, 73, 104
  high-chrome, low-calcium, 27–28, 69
  as indicator, 27–28
  in North America, 76, 77, 78, 81
  purple, 77
  pyrope, 78, 107
Gemprint, 198
Gemprint Corporation of Toronto, 198
Gems, 222, 236, 247
  Braganza stone, 17–19
  in India trade, 228
  storied, 17, 148
Gemstone diamonds, 239
Geological Survey of Canada, 70, 81
Geologists, 57, 60, 69, 70, 81, 93, 103
  Barrens strike, 97–99, 100, 101, 105, 107, 110
  De Beers, 71
  South Africa, 119–20
Geology, diamond, 28, 31, 32, 75, 80, 98
Geophysical reflection, 61, 62
Geophysical signature, 61
Geophysics, 77, 99
  airborne, 62, 63, 81, 83, 99–100, 109
Gerety, Frances, 140–41
Giglio, Luigi, 4–5, 6, 8, 9, 10, 12, 20–21
  diamond inventory, 14–15
Gimenez, Susana, 150
Glaciers, 69, 72, 73, 76, 79, 80
Gletz, 13, 206, 213
Global positioning system (GPS), 90, 92
Global Witness, 187–90
Gloworm Lake, 89, 91, 93, 94, 97
Goldberg, Whoopi, 143

Goldberg, William, 11, 202–4, 203*f*, 204, 205, 207, 208–10, 212, 213, 214, 221–22
Golden ADA, 175–80
Golden Jubilee, 221
Gooch, Charmian, 189
Gope pipe, 67–68
Government diamond valuator (GDV), 124, 125
Graphite, 22, 26, 252
Great Slave Lake, 77, 95
Green, Ben, 204, 208–10
Greenland, 119
Griqua people, 33, 34, 246
Griqualand West, 36
Grose, Sir Alan, 163–64, 168
Guillot, Cadet, 155, 156, 157
Guarantees of origin, 249–50
Guinea, 192
Gurney, John, 27–30, 31, 57, 64–65, 65*f*, 67, 69, 72, 73, 78, 80, 98, 251
Gwalior, rajah of, 145

H

Hain, Peter, 185
Hall, Tony, 190
Harris, David, 38, 53
Harry Winston (co.), 10, 11
Harzburgite, 30
Hearne, Samuel, 74–75
Hindson, Bob, 110, 111, 112
Hindustan Diamond Company, 135
Hirschhorn, Fritz, 48, 50–51, 53
Hollywood, 142–44
Hoover, J. Edgar, 158
Hope, Henry Philip, 157
Hope, Lord Henry Francis, 157–58
Hope Diamond, 144, 157–58
Hopkins, Robin, 105–7, 108, 109, 110
Hudson Bay, 73, 74, 77, 246
Hudson's Bay Company, 74
Hunt and Roskell, 34

I

I. Hennig & Son, 131
Ilmenite, 30–31, 67, 77
Immendorf, Jack, 178
Inclusions, 132, 147, 213
India, 156
  cutters/cutting, 147, 154, 227, 229, 230, 231, 236
  diamond business in, 32–33, 222
  diamond production, 37, 144
  polishing in, 224, 227, 228–36, 238
  sales of finished goods, 223–24
India trade, 222, 224, 228, 231, 233, 236
  center of, 238–40
  crisis in, 234–35
  small diamonds in, 236
Indians, 74, 75
Indicators (diamond indicator minerals), 27, 29, 30–31, 56–57, 58, 64, 70
  in Barrens, 76, 90, 104
  in North America, 77, 79, 80, 81–82
  in sample from Lac de Gras, 107, 108, 110
Industrial diamonds, 228, 236
Inspection, independent, 250
International Corona, 77
International Diamond Manufacturers Association, 194
Isotopes, 196–97

J

Jennings, Chris, 57–58, 60–68, 62*f*, 72, 80*f*, 103, 184
  diamond hunt in South Africa, 119–21

diamond hunt in North America, 77, 78, 81, 87–92, 96
head of SouthernEra, 250
Jewel (the), 200
Centenary Diamond, 220
hypothetical, 206
new order of, 239–40
revolution in definition of, 222
in the rough, 208
Jewelers
stealing diamonds, 159–60
Jewelry, 137, 139–44
market for, 29
small diamonds in, 227
Jewelry stores, 136–37
Joel, Solly, 49, 50, 52
Johannesburg, 61, 62, 63, 72, 115, 134, 135
Johannesburg Stock Exchange, 29, 115–16
Juniors, 85–86, 87, 95, 103, 104, 119, 120, 184
Jwaneng, 58, 65, 66–67, 171, 236

K

Kaapvaal Craton, 33, 55–56
Kabila, Laurent, 186–87
Kalahari Desert, 33, 55–58, 61, 62
diamond hunt in, 62–68
Karp Diamonds, 229
Kelowna, 76, 78, 80
Keough, Leni, 77–78, 90, 97, 98, 251
Kerfing, 217
Kimberley, South Africa, 24, 36, 39, 41–42, 45, 48, 215
diamonds from, 59
early digs at, 40*f*
Kimberley Central Diamond Mining Company, 41–43
Kimberley pipe, 39–40
Kimberlite, 24, 25, 26, 27, 39, 76, 78–79, 80, 99
in Canada, 69
erupting, 70
extracting, to count diamonds, 103
extracting ore sample from, 66*f*
from Lac de Gras, 106–7, 108, 110
samples of, 64, 65, 67, 83, 104
South Africa, 29
Kimberlite fissure: Springbok Flats, 119–20
Kimberlite pipe, 24, 25*f*, 71
in clusters, 30–31
Kleinsee, 58, 160, 170
Knot(s), 206, 208
Koh-i-Noor, 139, 144–48, 153, 229, 252
Kollur, 144
Kopp, Quentin, 178
Kozlenok, Andrei, 174–80

L

Lac de Gras, 76, 77, 78, 81, 93, 99, 101, 103–4, 242, 243
camps at, 246–47
diamonds in, 102, 244
drilling in, 104–12, 106*f*
strike, 110–12, 113
Lac du Sauvage, 91, 97–98, 99, 101
Lamont, Gavin, 55–58, 61, 62, 63, 66
Laurentide ice sheet, 73
Letlhakane, 58, 65
Liberia, 182, 187, 192
Lifshitz, Mervin, 7–9, 12–13
Lobatse, 55, 57, 60, 61, 66
London, 44, 135
diamond quarter, 77–8, 122*f*
financial district, 121
Hatton Garden, 194
London diamond syndicate, 47, 48, 50, 123
Oppenheimer destroyed, 53
London stockpile, 136, 193, 248
Louis XIV, 139, 156
Louis XV, 154
Louis XVI, 153–54

Luanda, 183, 184, 185, 189, 193
Luo diamond ground, 185
LVMH, 137

M

McConnell Range, 71, 72
McLean, Evalyn Walsh, 158
Mackenzie River, 71, 72, 73
Magnetism, 99–100
Maha, Princess, 221
Maharaja of Nawanagar, 221
Maldar, Kishor, 229
Mandela, Nelson, 115
Manitoba, 251
Mantle, 24, 25, 26, 30, 33
Maria Luisa, Queen, 156
Marks, Manfred, 58
Marsfontein pipe, 120–21, 250
Mehta, Arshad, 232
Mehta, Arun, 231–32, 234, 235 239
Mehta, Dilip, 231–32, 233, 235
Mehta, Ramniklal, 230–31
Mehta, Russell, 223, 224–26, 227, 239
Mehta family, 230–36, 239
Menell, Brian, 19
Meteorites, 23
Microdiamonds, 78, 104, 107, 108
Microprobe, 27, 78
Middle empire, 32–59
  borders gone, 118
Miette, Paul, 155, 156
Minas Gerais, Brazil, 1, 2, 14–15, 17, 18
Mineral exploration
  in Canada, 85–86
Mineral rights
  in Canada, 79
Minerals, 26–27, 69
  in diamond formation, 27, 30
  see also Indicators (diamond indicator minerals)
Mines/mining, 236
  distinct productions of gems, 196
  gem-grade stones, 236
  India, 229–30
  new, 55
  South Africa, 50, 51
  stealing from, 161–62, 170, 171, 251
  technique for finding, 31
MLG 2.5–4, 132
Mobil Oil, 76
Monopros, 71
Moore, Rory, 78–79, 80, 81, 98, 251
Morgan, J. P., 49, 51
Mostert, Frikkie, 161
Mvelaphanda Diamond, 116

N

N. W. Ayer & Son, 140, 141
Nadir Shah, 139, 252
Namaqualand, 52, 160–61, 163, 166, 169
Namdeb Diamond Corporation, 160, 161, 162, 163–64, 168–69
Named diamonds, 17, 148–49
Namibia, 50, 51, 58, 160, 161, 164, 165, 186
Namibian Diamond Corporation (Namco), 117
National Union for the Total Independence of Angola (UNITA), 183–86, 189, 190, 191, 192, 195, 249
Native peoples (Canada), 243–44, 246
Navigator Exploration Corp., 251
Near gems, 228
New York City, 10, 19, 222, 235
NGOs (nongovernmental organizations), 187, 248, 249
Niekerk, Schalk van, 33–34
Nkuhlu, Wiseman, 116
Nobels, Claude, 124–25
North America, diamond hunt in, 68–83

Northwest Territories, 76, 77, 102, 197–98
Northwestern Canada (map), 75*f*
Nugget effect, 64

O

Ocean Diamond Mining (ODM), 117
Ocean mining, 164–69
Ontario, 69, 119, 251
Oppenheimer, Bernard, 47
Oppenheimer, Sir Ernest, 47, 48–53, 49*f*, 114, 123, 125, 140
  elected chairman of De Beers, 53
Oppenheimer, Harry, 68, 72, 126, 127, 134, 148, 182
Oppenheimer, Louis, 48, 51
Oppenheimer, Nicky, 117, 123, 131–32, 248
  on Centenary Diamond, 220
  and war diamonds, 194
Oppenheimer, Otto, 48
Oppenheimer, Philip, 125
Oppenheimer family, 29, 114, 118, 123, 247
  at head of diamond world, 116
Orange Free State, 35, 36
Orange River, 33, 116, 117–18, 160, 166, 251
Orapa diamond pipe, 58, 60, 61, 62, 65
  geophysical signature, 61–62, 63
Orapa rough, 196–97
Overburden, 56, 64, 105, 106

P

Paltrow, Gwyneth, 143–44
Parker, Stafford, 35, 36
Partnership Africa Canada, 192
Patos de Minas, 2, 3, 14, 15, 19
Penny stock, 86
Peridotite, 30
Perón, Eva, 149
Perth, 171, 172
Pigeon scheme, 162–63
Pink diamonds, 5, 153, 171, 172, 184, 236
  false, 3
  large, 1–21, 241
  prices, 14
Pipe(s), 22–23, 25–27, 32, 33
  barren, 64
  in Barrens, 98, 99, 102–3, 104, 113
  clusters of, 30–31, 63
  exploration of, 64
  finding, 26–27
  geophysical reflection of, 62
  highest-grade cluster of, 112
  in Kalahari, 64–65
  under Lac de Gras, 247
  North America, 69, 70–71, 79–80, 83
  predicting diamond grade in, 67
  Siberian Craton, 54
  South Africa, 27, 55–56
  technique for finding, 28–30, 31
Pitau, 154
Point Lake, 82–83, 84, 85
Point Lake pipe, 99, 250
Polariscope, 210
Polished goods, demand for, 232
Polishing, 194–98, 208
  advances in, 222
  carats lost in, 19
  Centenary Diamond, 219–20
  in India, 223–24, 227, 228–36, 238
  large pink, 12–14
Popular Movement for the Liberation of Angola (MPLA), 183–84, 186, 189
Port Nolloth, 169–70
Portugal/Portuguese, 18, 170
Post, Marjorie Merriweather, 149
Prasad Chambers, 238–40
Premier mine, 45–47, 58, 214, 217
Premier Rose, 11

Presidente Varga (diamond), 5
Price(s), 9, 54
  colored diamonds, 14
  control of, 29
  cut and, 211–12
  De Beers controlled, 53
  India, 232
  manipulation of, 113, 135–36
  pinks, 14, 20
  rough, 44
  stability, 45
Proportion, importance of, 211*f*
Provenance, 148, 150, 192
  lacking, 174
  Queen of Holland diamond, 221
Pumpkin diamond, 11, 209
Pyropes, 30, 31, 78, 107

Q

Queen of Holland diamond, 221–22

R

Rae, 243, 244
Ralfe, Gary, 199, 249–50
Ranjit Singh, 145–46, 148
Rapaport, Martin, 193–94
Recovery process, 167–68
Rhodes, Cecil, 37, 41–44, 42*f*, 51, 53
Rio Abaete, 1–2, 5, 17–18, 241, 252
Rio Santo Antonio do Brito, 5–6
Rio Tinto, 94–95, 103, 104–5, 113, 236, 238, 247, 250
  Diavik site, 242–43
  and Dogrib, 245–46
  drilling protocols, 109
Rivers, diamonds from, 5–6, 14, 19, 22, 33, 57, 58
  see also Alluvials
Rocky Mountains, 70, 71
Roddan, Lindsay, 171–72, 173
Rossman, George, 196–97, 250
Rosy Blue, 233–40, 234*f*
Rothschilds, 42, 49, 53
Rough, 10, 16, 112, 114, 252
  African, 192
  from Angola, 186
  assortment of, 13*f*
  from Botswana, 58
  in Brazil, 5–6
  categories of, 124–25
  in Charterhouse Street, 121–22
  controlling supply of, 55, 135–36
  De Beers, 122, 123
  De Beers control of, 45, 114, 136
  De Beers sales of, 84–85
  demand for, 29
  determining origin of, 192, 195–97, 250
  diamantaires' passion for, 12
  grading, 132
  illicit, 183, 187, 192, 249
  in India trade, 230, 231, 232, 233, 234, 235, 237, 238
  origins of, 248–49
  potential and risk in, 12
  price for, 44
  sales, 139
  sales from mines, 137
  seabed mining, 165
  secondary trade in, 233
  sorting, 123–28
  speculation in, 235–36
  stealing, 251
  stolen from De Beers, 180–81
  surcharge on, 235–36
Rough market, 54
  De Beers share of, 247, 248
Rough Trade, A, 187–88, 189
Rupert, Johan, 117, 118
Rupert family, 117, 118
Russia, 119, 122, 134, 135, 183, 247
  diamond thefts, 181
  State Treasury, 175, 179
Russian rough
  selling to American market, 174–80

Russian technicals, 237
Rwanda, 186

S

SADE (Scale Automatic Diamond Electronic), 124
"Salting," 28
Salzman, Ned, 209–13
Sampling, 25, 56, 57
  in Barrens, 98, 99, 100, 104
  in North America, 69–70, 77, 78
San Francisco, 85, 174, 175
  Russian rough in, 175, 176, 177, 180
Saskatchewan, 119
Savimbi, Jonas, 184
Saxondrift, 251
Schiller, Ed, 81–83, 96, 250
Schönwandt, Bjarke, 97, 98
Schuler, Gary, 149
Second Mazarin, 156
Secrecy, 69, 80, 214, 250
  commercial, 196
  De Beers, 28
Selection Trust, 125–27
Selling mixtures, 131
Seniors, 119, 251
Sexwale, Tokyo, 115–18, 116*f*, 251
Shagirian, Ashot, 175, 180
Shagirian, David, 175, 180
Shah, Arvind, 225–26, 227
Shah, Parag, 229
Shapes, 12, 124, 127, 132
  Centenary Diamond, 218–19
Shared-electron bond, 24
Siberian Craton, 54
Sierra Leone, 134, 190
  rough from, 125, 126, 197
  war diamonds, 182–83, 195, 199, 248
Sight week, 132–33
Sightholders, 122, 123, 131, 133, 135, 203, 235
  Indian, 237
Sights, 122–23, 199, 233
  five-week cycle, 122–24, 131–33
Sillitoe, Sir Percy, 182
Single-channel marketing, 53
Size, 3–4, 124, 127
Slave Craton, 72, 76–77, 81
  magnetism, 99–100
Small diamonds
  have value, 239–40
  India trade, 228, 236
Smithsonian Institution, 149, 158
Smuggling, 182, 190–92
Snap Lake, 119, 246
Soil analysis, 196–97
Sorters, 121, 127, 128, 231
South Africa, 34–37, 48, 50–51, 54, 84, 98, 215, 230
  cartel in, 114–21
  criminal syndicates from, 171
  diamonds from 44, 58–59
  government diamond valuator, 124
  kimberlites in, 29
South African Police
  Gold and Diamond Branch, 170
South West Africa, 50
South West Africa People's Organization (SWAPO), 164
Southern Africa (map), 52*f*
SouthernEra Resources, 96, 119–21, 250
Southey, Sir Richard, 34
Soviet Union, 54–55
Spectroscopic signature of diamond, 22
Speculation, 12, 235–36
Speculative capital, 118–19
Sperrgebeit, 160–61
Springbok Flats, 119–20
Star of South Africa, 34
Stealing, 128–30, 159–64, 168–71, 180–81, 251
  French Blue, 153–58

Stephenson, John, 110–11
Stolen diamonds
  in legitimate trade, 174–81
Stone(s), 132, 200
  grain of, 204
  modeling, 209–10
Streeter, Edwin, 153
Superior Oil, 63, 68, 76, 82
  diamond exploration in North America, 68–69, 72, 73
Supply, control of, 29, 44, 63
Surat, 225, 227, 232, 240
  cutters in, 224
  diamond trading in, 228*f*
  polishing factories in, 227, 228–29, 230
  switching market in, 227–28
Switching, 128–29, 227–28
Switzerland, 249
Syndicates
  competing, 51, 52, 53
  criminal, 163–64, 171, 192
  see also London diamond sydicate

T

Tanzania, 58
Tavernier, Jean-Baptiste, 144, 154, 229
Teakle, Simon, 150–53, 151*f*
Technology
  cutting, 217
  in diamond wars, 194–98
Teixeira, Antonio, 185
Tel Aviv, 19, 54, 84, 122, 135, 203, 249
  cutters/cutting, 222, 224, 229, 232
  speculation in rough, 235
  war diamonds in, 183, 187, 192
Thatcher, Margaret, 218
Thomas, Eira, 93–94, 94*f*, 96, 97–98, 101–2, 103–12, 111*f*, 242, 247, 251
Thomas, Grenville, 86–87, 88–89, 88*f*, 91, 93, 94, 99, 100, 103, 104, 107, 251
  and Lac de Gras strike, 112
Thompson, Julian Ogilvie, 215
Thor (dog), 97, 98, 101–2
Tiffany & Co., 143, 211, 247
Tli Kwi Cho, 104, 112
Tolkowsky, Gabi, 212, 214–21, 221*f*, 222, 251
Tolkowsky, Isidore, 216
Tolkowsky, Marcel, 216
Tolkowsky, Maurice, 216
Toronto Stock Exchange, 93, 96
Tower of London, 46, 144–45
Trans Hex Group, 117–18, 169, 251
Treaty of Lahore, 145–46

U

Uganda, 186
Ukraine, 119
UNITA
  see National Union for the Total Independence of Angola (UNITA)
United Nations, 183, 184, 185, 188, 190, 193
  Security Council, 190, 249
United States, 183, 189, 190, 199–200
  BHP in, 85
  diamond market, 113, 136, 174–80
United States Congress, 250
United States Department of Justice, 85
United States Geological Survey, 69, 187
Universe, diamonds in, 22–23
University of Cape Town
  kimberlite lab, 78
Unnamed Brown, 217, 220–21
Uplift theory, 58
Uruguay, 119

V

Vaal River, 33, 34, 35, 36
Valuation, 4, 55, 125–27, 237
  large pink, 6–10, 13–14, 20
  provenance in, 150
Van Cleef & Arpels, 143, 149
Van der Westhuizen, Hennie, 119
Vancouver Stock Exchange, 76
Venetia, 58
Victoria, Queen, 146, 147–48
Volcanic eruptions, 22–23, 24–26, 40

W

Wake-Walker, Richard, 7, 8, 9, 12, 133–35, 181
Waldman, Mike, 97
Walsh, Mary, 141–42
War diamonds, 181, 182–200, 248, 249–50
Warranty chain, 194, 248–49
Waterfall, 152
Weight
  cutting for, 211
  lost in polishing, 12
  rounding down, 129–30
Wells, F. G., 45
Wernher, Beit and Company, 48
West Africa, 52, 125, 126
Western Australia State Bureau of Criminal Intelligence, 172
Wharton-Tigar, Edward, 125–27
White diamonds, 15, 208–9, 221, 227
  cutting, 211
  Lac de Gras, 244
White House Office of Science and Technology, 196
Whiteout, 92
William Goldberg Diamond Corporation, 201–2
Winans, Louis, 153
Window, 10, 204–5
Winspear Diamonds, 119
Winston, Harry, 143–44, 148, 149, 158, 209
Winston, Ronald, 10
Winter road, 95, 104, 105*f*, 242, 243
  trucks waiting to travel, 246, 246*f*
Wodehouse, Sir Philip, 34
World Diamond Council, 194
World Federation of Diamond Bourses, 194
WWW International Diamond Consultants, 129, 134
Wyndham, Charles, 129–30, 134

X

X ray scanner, 163, 164

Y

Yamba Lake, 96
Yearsley, Alex, 189
Yellow diamond(s), 202–3, 204–14, 205*f*
Yellow ground, 39–40
Yellowknife, 76, 78, 79, 88, 89, 90, 93, 95, 97, 101, 102, 103, 107, 112, 242
  polishing factories in, 197
  processing mill at, 104
Yeltsin, Boris, 175, 179
Yetwene (mine site), 185, 186
Yield, polished, 12
  large pink, 13–14

Z

Zhirov, Viktor, 177–80
Zimbabwe, 186
Zoe, John B., 243–45

Yellowknife, Canada
NORTH AMERICA
New York City, U.S.A.
Pacific Ocean
Atlantic
SOUTH AMERICA
Minas Gerais, Brazil
N
0 Miles
2000
4000
0 Kilometers
4000
© 2001 Jeffrey L. Ward